D0707752

CUTTHROAT

CUTTHROAT

High Stakes & Killer Moves
on the
Electronic Frontier

STEPHEN KEATING

JOHNSON BOOKS

Boulder

Copyright © 1999 Stephen Keating

All rights reserved. No part of this publication may be reproduced or transmitted in any form or by any means, electronic or mechanical, including photocopy, recording, or any information storage and retrieval system, without permission in writing from the publisher.

Published in the United States by Johnson Books, a division of Johnson Publishing Company, 1880 South 57th Court, Boulder, Colorado 80301. E-mail: books@jpcolorado.com

9 8 7 6 5 4 3 2 1

Cover design: Debra B. Topping
Cover illustration: James Lee Roth

Library of Congress Cataloging-in-Publication Data
Keating, Stephen.
    Cutthroat: high stakes and killer moves on the electronic
frontier / Stephen Keating.
        p.   cm.
Includes bibliographical references and index.
    ISBN 1-55566-252-8 (alk. paper).—ISBN 1-55566-248-X (pbk.:
alk. paper)
    1. Telecommunication—United States—Marketing.
2. Telecommunication policy—United States.   3. Competition—United
States.   4. Monopolies—United States.   I. Title.
HE7775.K4   1999
384'.0973—dc21   99-34764
                                                                                   CIP

Printed in the United States by
Johnson Printing
1880 South 57th Court
Boulder, Colorado 80301

 Printed on recycled paper with soy ink

It is a world in which competition is lauded as
the basic axiom and guiding principle,
yet *too much* competition is condemned as *cutthroat.*

— Alan Greenspan, 1961

CONTENTS

THE GAME

The game is going on right now, all over the world. It requires no allies, just alliances. No team loyalties, just self-interest. Strategies form and dissolve with every play. A helping hand may bring a knife in the back. In the next round, the players switch sides and do it again.

Cutthroat is the name of the game on the electronic frontier.

Cable cowboys once lassoed broadcast TV signals into rural towns, then sought to strangle the broadcasters. Satellite TV aims a laser at both. Phone companies must connect the calls of competitors or die of isolation. The Internet is a vast, chaotic network that owes its nature to print, phone, broadcast, cable, satellites and computers, yet is something else entirely that may devour all of it.

Electronic media is an interlocked, trillion-dollar, global game that feeds on our insatiable desire to know, communicate and be entertained. The game is guided by technology, big money, regulation, demand and the artful savagery of media chiefs.

They need each other, as partners and suppliers. Yet they bleed each other as rivals and takeover targets. It all depends on the day, the deal and the leverage to get it done.

They're playing cutthroat in your home.

With such thievery and cross-purposes built into the game, who wins and who loses? What are the stakes? The rules?

The essence of the game is illustrated by one play.

In February of 1997, Rupert Murdoch, the Australian media pirate, and Charlie Ergen, a wily American hustler, announced plans to launch a killer satellite TV service nicknamed Deathstar. It lit up the sky, then fell to earth, undone by a cable mastermind named John Malone.

How these players and others came to be, the way that they clashed and what happened next is a view to the future of the electronic frontier.

ONE

THE CABLE
GANG MOURNS

O UTSIDE THE CHURCH in suburban Denver were nine limousines, eight white and one black, disgorging men in dark
suits. In a red-railed enclosure nearby, an Arabian stallion stood
bridled in silver and leather, nibbling on a bale of hay.

Inside the First Church of the Nazarene, 1,000 people took their
seats on hardwood pews, facing a stage covered with 750 white
roses. On an easel was a large photo of a wiry Oklahoman named
Bob John Magness, smiling, with a whistlin' gap between his teeth
and a crow's beak of a nose. His cowboy hat hung from a corner of
the picture frame. A Boy Scout color guard marched in with three
flags, representing the United States of America, the state of Colorado, and the blue-and-purple corporate stripes of Tele-Communications, Inc., the world's most powerful cable TV company,
known simply as TCI.

Beyond those in the church, including Ted Turner and other
media executives from around the country, few people knew Magness. His death of cancer at age 72 in November of 1996, and life up
to that point, occasioned little notice outside the industry. But
through timing, perseverance and dumb luck, Magness had risen
over four decades to head a colossus.

One in four cable subscribers signed a check each month to TCI

1

or its affiliates. TCI owned pieces of many networks, ranging from CNN to the Discovery Channel to Black Entertainment Television to Fox Sports. Internationally, TCI operated in England, Japan and Argentina. Along the way, Magness accumulated a billion-dollar fortune, part of which he spent on raising his prized Arabian horses and collecting western art.

Born in 1924, Magness served as an infantryman in World War II, graduated from Oklahoma State College and tripped into the cable business in 1948 after selling cottonseed in Texas. He liked to do deals on a handshake, didn't say much and didn't have to. He was a taciturn westerner who headed the only one of six major cable families not based on the East Coast.

There was Cox in Atlanta, Comcast in Philadelphia, Cablevision on Long Island, New York, and Continental in Boston. They were known collectively as the four Cs, all tough in their own ways. There was also Time Warner, Inc., based in Manhattan, a bruising conglomerate.

But the desperado of the bunch, the universally acknowledged badass of the cable business, was TCI in Denver. It was the most in-fluential and the most feared because it was the most mercenary. TCI did a deal a week and treated its local cable monopolies like rickety cash machines, skimping on upkeep and piling on debt to buy even more systems.

"I'd have to think harder to quit doing acquisitions than I would to keep doing them," Magness once said, his folksy manner belying a voracious appetite for expansion.

Cable had grown up in places like Astoria, Oregon; Memphis, Texas; Mankato, Minnesota; and Mahanoy City, Pennsylvania. They were communities isolated from the national conversation that was television. Magness and other cable cowboys shimmied up poles and caught splinters hanging wire, collecting a few dollars each month from every customer they could hook up in 1950s rural America.

Cable began by carrying the signals of independent TV stations and the Big Three: ABC, CBS and NBC. The broadcasters intended to dominate TV forever. But cable relied on the broadcasters to grow up, then blew past them. With cable-only networks like HBO, ESPN and MTV as a draw, 40 million new subscribers signed up in the 1980s, making cable almost as much of a household utility as phone, natural gas and electric service.

By 1990, TCI's revenues were more than the Big Three broadcasters combined.

From rural roots, the cable industry evolved into a loosely grouped gang with internal discord, but with even greater common foes: obstructionist broadcasters, the Hollywood studio syndicate, the heavyweight Baby Bells, an antagonistic press, a regulatory Congress, parochial town councils and stealthy cable pirates. The cable chiefs bought, sold and swapped systems like tribal lords. They carved up urban markets and grew rich from owning stock in monopolies built from the ground up. Ten men on the *Forbes* list of the 400 richest Americans made their money in cable and programming, though few of the names besides Ted Turner were well-known.

The cable gang had only one great fear: competition.

By the time of Magness's death, cable's boom years were long gone. The industry was spending billions of dollars to rebuild its aging coaxial networks into a pipeline to carry Internet and phone traffic—a market the Baby Bell phone companies wanted for their own. Meanwhile, the cash flow funding the cable pipeline construction was under siege. The federal government had capped cable rates in 1992. Programming costs were rising.

From the sky came a killer: satellite TV. Companies like DirecTV and EchoStar offered nearly 200 channels of diamond-sharp TV pictures and sound, delivered by satellite to 18-inch dishes attached to homes. Cable subscribers angry at rate hikes and poor service had an alternative. Millions grabbed it. DirecTV became the fastest-selling consumer electronics product in history, faster than the VCR, the fax machine and the TV itself.

The cable gang was in a space-based pincer. DirecTV, backed by the deep pockets of General Motors' Hughes Electronics, targeted cable's highest-revenue customers with exclusive sports and an array of pay-per-view movies. EchoStar trolled the low end by basically giving away satellite systems to capture monthly programming fees. Struggling to remain competitive was the cable gang's own satellite TV service named Primestar.

So, as Ted Turner and his cable brethren filed into church on that crisp November day in 1996, they grieved for one of their own, a godfather of the cable industry, Bob Magness. But they might well have mourned the death of another entity that had brought them power and riches beyond their wildest dreams: monopoly.

Deal-driven and debt-laden, TCI faced particular pressure from satellite TV's airborne assault. Even as it added more and more subscribers through acquisitions, TCI lost tens of thousands of basic subscribers and nearly one million premium subscriptions to lucrative pay channels like HBO in 1996. TCI's $15 billion debt was six times its cash flow, a precarious leverage ratio. Investor and banker confidence in the company's management eroded. Its stock price slid to a sickly $12.

Two weeks after the Magness funeral, just before Christmas, TCI would cut 2,500 of 38,500 jobs, slice executive salaries by 20 percent and put its four corporate jets up for sale. The Magness estate's big block of TCI stock was up for grabs, while takeover vultures, including Microsoft chief Bill Gates, circled overhead.

John Charles Custer Malone, the man who had the most to lose from this troubling convergence of events, sat in a front pew of the church, awaiting his turn to memorialize Magness—his mentor, business partner and close friend through three decades. Magness was TCI's founder and chairman, but the cable company's financing schemes and aloof corporate culture flowed from the mercurial Malone.

Malone was a problem solver, not a glad-hander. He could be gracious and charming, or cold and cutting, in the same tone of voice. He took the long view in his business and personal life. He had been TCI's chief for 23 years and married for 33 years to his hometown sweetheart, Leslie, with whom he had raised a son and a daughter. He was an engineer in the purest sense of the word, a man at ease building things, whether it was a cable television empire or Legos on the floor of his office with a grandchild.

His handsome face was cut into sharp planes, with a straight nose and thin lips. His dark hair, going gray, was short and neatly parted on the right. His eyes were deep and watchful, like a shark measuring its prey. At 55, Malone had the wealth, position, power and perks commensurate with running a multibillion-dollar media conglomerate.

Malone shielded his private life, even as he dominated an industry whose purpose and profit are founded on exposure. Few people had ever been to his house or met his wife, an art collector and charity volunteer whose fear of flying dictated that Malone drive them across country to their vacation home in Boothbay, Maine. While expressing fealty to his wife, Malone was absolutely ferocious in his business life. Whether TCI subscribers were pleased with their cable service or not, whether TCI was seen as a good corporate citizen or not, had little to do with Malone's singular mission.

"TCI is purely and simply a wealth creator," he said many times. In this, he embodied the raw capitalist spirit described by Scottish economist Adam Smith in 1776, who wrote, "By directing that industry in such a manner as its produce may be of the greatest value, he intends only his own gain, and he is in this, as in many other cases, led by an invisible hand to promote an end which was no part of his intention. … By pursuing his own interest he frequently promotes that of the society more effectually than when he really intends to promote it."

Thus was the conundrum behind Malone's piercing hazel eyes in the winter of 1996. Was he a deal-making visionary helping to build a communications railroad for the twenty-first century? Or a heartless monopolist constructing a debt-laden pyramid for the enrichment of himself and his cable cronies? Or both?

"I would call myself a corporate architect, a systems engineer," he once said. "I like to play the strategic games, a whole lot of moves into the future."

When things got bad, and they were about to get very bad in the months following Magness's death, Malone relied on his prized asset: his mainframe brain. With a doctorate in operations research, Dr. Malone—as he was sometimes known—often worked backward from the answer he desired to define the problem. This worked well with Turner and Rupert Murdoch, two gut-driven moguls in awe of Malone's calculating mind. This tactic was of less use with Gates, who shared Malone's hyper-rational worldview. As software and cable technologies converged in the promise of a new set-top box to give the dumb TV a computer brain, Gates and Malone acted at times like two scorpions trapped in a bottle of their own making, seeking an advantage that could only come at the other's expense.

In public, Malone was alternately imperious and shy, with a wicked sense of humor. He usually said what he thought, not what he felt. But on the occasion of his mentor's funeral, Malone revealed a deep filial bond.

"I've probably given 1,000 speeches in the past 30 years," Malone began, standing at the lectern, the flag-draped coffin between himself and his listeners. "This is the toughest one for me."

Magness had hired Malone when the younger man was president of Jerrold Electronics, a Pennsylvania company that supplied remote controls and cable equipment.

"When I finally got to Denver," Malone told the mourners, "I found that Bob had rearranged our offices so that we were back to back. He had his office. I had mine. There was a pathway in between. For 25 years, that's been the configuration wherever we've gone. I've always had Bob protecting my back."

Malone, the Yankee engineer, and Magness, the sharp-eyed cowboy, were an odd couple and inseparable. They rarely bickered and even stayed in the same hotel rooms on the road in the 1970s as they staved off financial ruin at TCI. Magness was the wise elder and Malone the muscle who would attempt to control the cable industry as John D. Rockefeller did the oil business a century before. At its height, Rockefeller's Standard Oil Trust controlled 90 percent of the country's oil production, a lucrative clamp on both the pipeline and the oil that flowed through it, effectively fixing prices.

Malone would attempt a similar feat with the cable pipe and the programming that flowed through it. This was seen as arrogance by competitors, with disdain by critics and as cause for alarm by regulators. As with Rockefeller, the fear of Malone was born from his particular genius. This did not just mean very smart, which he was, or ruthless, which he could be. It meant that Malone knew more about the economics and technology of his chosen field—and how to work these to his advantage—than anyone else. Rockefeller did not start out the biggest, nor the richest, nor the smartest. He achieved great power by accepting without emotion how the world works and by bringing an analytical mind to bear on the underlying problem, which is where to apply leverage to achieve the maximum effect.

Leverage rules. That Malone could see forward and calculate backward helped him apply this lesson earlier, more consistently and more fruitfully, than his colleagues and rivals. Thus was his reputation built, his counsel valued, his favors sought and his retributions avoided.

Malone entered, then dominated, a business that was a modern-day gusher, with the gloss of entertainment, the economics of a utility and the promise of a monopoly. Malone owed much of it to Magness.

"I came to love Bob," he told the mourners. "A man of enormous virtues. Quiet, but deep, solid, loyal. Brilliant in many ways, self-effacing. We went through a lot of experiences together. We were partners, he was my mentor. In many ways, he was my father."

Malone addressed a congregation that had done many deals together, fought many battles and kept many secrets. Al Gore, as a rising U.S. senator from Tennessee, had described them as the "Cable Cosa Nostra, the family that controls the distribution," with Malone its unscrupulous head. Malone despised Gore's politics and grandstanding. Yet, Malone's oration at Magness's funeral echoed Gore's criticism and elicited some knowing chuckles.

"We were frequently unindicted co-conspirators," Malone said of himself and Magness. "We built a company. I joined a family."

A prominent member of the cable family, and the next to eulogize Magness, was Turner, the Atlanta media maverick and vice chairman of Time Warner Inc. Lithium treatments and marriage to Jane Fonda had somewhat tamed Turner, though there was no restraining "The Mouth of the South." There was also no denying his peculiar stamp on American culture. His Atlanta Superstation, CNN, the Cartoon Network and Turner Classic Movies supplied much of the video buzz that flowed through the lines of TCI, Time Warner and much smaller cable companies into tens of millions of TV sets worldwide.

In the late 1980s, however, Turner's company was on the brink after a high-priced deal to buy the MGM studio and film library. TCI and other members of the cable gang kept Turner afloat, while tying his hands so that any big-ticket purchase needed approval by a new board of directors, which included Malone.

"I've got him right where I want him," Malone reportedly said of Turner.

Turner was a little more earthy when he described Time Warner's restrictions on his ability to buy a broadcast network or some other media toy.

"I'm being clitorized," said Turner.

Still, Turner owed much of his success, and eventual Time Warner riches, to the intercession of Malone and Magness. It was fitting that he speak at Magness's funeral, though his comments were as much a self-tribute as a memorial. Such a ham was Turner that some of the mourners felt the corners of their mouths lift involuntarily, tickled by his deep-fried stream of consciousness, delivered in a croaking singsong.

"I hope there's this many people at my funeral," said Turner, who at one point blew his nose into a white handkerchief, then stared briefly at the result.

"I always used to say, 'You build the wires and we'll make 'em sing,'" Turner said of Magness and Malone. "And it really worked out. Then we needed hundreds of millions of dollars and Bob and John came forward and became equity partners in the company." Turner credited Magness with not only saving his company but introducing him to the West, great swaths of which Turner subsequently bought in New Mexico and Montana, becoming one of the nation's largest individual landholders with one million acres.

"Those mountains that run all the way from Canada to, I don't know, Mexico," said Turner. "And I'm trying to buy as much of it as I can."

Turner said that Bob and Sharon—referring to Magness's second wife, who sat solemnly in a front pew—had come to visit him and Jane Fonda at their ranch in Montana, a state where Magness began building his cable empire in the 1950s. Turner and Magness stopped by a microwave transmission tower that Magness once serviced during a snowstorm.

"When we first met, I didn't have two nickels to rub together," said Turner. "Before it was over, we were talking about hundreds of millions, you know. So we really were about as successful as you could possibly be."

After the funeral service, pallbearers including Malone, Turner and Magness's two sons by his first marriage, Kim and Gary, carried the casket containing the fallen patriarch to a waiting hearse. The mourners poured out of the church, having paid their respects to a man whose company, for better and worse, defined the U.S. cable industry.

One executive recalled a baseball playoff game in Denver the year before between the Colorado Rockies and Turner's Atlanta Braves. TV commentators spotted Turner and Malone in a lower-level box and wondered aloud "who the man sitting next to them is." It was Magness, of course, whose low profile hid much wheeling and dealing. His greatest coup was hiring Malone, whom Magness once described as "the smartest son-of-a-bitch I have ever met."

Few in the cable gang, particularly those at the funeral, would disagree. All of them had ties to TCI and most of them owed Malone and Magness some debt of loyalty. There was Glenn Jones, a

Colorado cable lawyer who bought his first cable system by bor-
rowing $400 against his Volkswagen in 1968. He went on to build
one of the nation's top ten cable companies and launched an educational channel that initially relied on TCI for distribution.

There was Ralph and Brian Roberts, the father-and-son team
who headed Comcast, based in Philadelphia. TCI and Comcast had
several profitable investments in common, yet there was a rivalry
between Brian Roberts and Malone.

There was Daniel Ritchie, the chancellor of the University of
Denver and a former chief of Westinghouse Broadcasting. A longtime Magness family friend, Ritchie was a co-executor of the Magness estate, which left $10 million to the university.

And there was Tim Robertson, son of televangelist Pat Robertson. In 1990, TCI helped the Robertsons finance a lucrative separation of their Family Channel from the company's religious ministry
based in Virginia Beach, Virginia.

Out of loyalty and self-interest, it was unheard of for these men
or any others within the cable industry to criticize TCI publicly,
even though the company's arrogance had given the industry a
black eye and helped lead to kneecapping rate regulation by the
federal government in 1992. Within the cable gang, testosterone,
shared experience, blood oaths and distrust of outsiders ruled the
day.

Malone was the kingpin.

He suspected, but could not foresee, the turmoil that was about
to engulf TCI and the cable industry, events that would send its
prospects plunging and call into question his own vaunted ability
to sell the electronic frontier to Wall Street, investors and the media.

One of Malone's duties, as TCI grew influential domestically and
overseas, was to negotiate with Murdoch, carving up the world in
their minds if not in reality. They attended to their mutual business
interests like heads of state. Murdoch's News Corp. was the supplier
of popular Fox networks to TCI and the cable industry. The two
men were partners in billion-dollar programming ventures in the
United States and satellite TV service in Latin America. In England,

however, they competed head-on. Murdoch's satellite TV service, British Sky Broadcasting, dominated. It served five million households and controlled key sports and movie rights. TeleWest, a company part-owned by TCI, was BSkyB's landline competitor, offering cable and phone service.

Murdoch and Malone played an endless, international game of cutthroat, partnering here, competing there and always playing on the edge. In early 1997, as Malone and the cable gang had their backs against a wall, Murdoch took aim. He had long wanted to control satellite distribution in the lucrative U.S. market, as he did in England. To advance his plan, Murdoch picked a partner right in Malone's backyard, a renegade named Charlie Ergen, who intended to put a satellite dish on every home and put John Malone and the cable gang out of business.

TWO

DEATHSTAR

I NTRODUCED BY horn-heavy instrumental jazz, Rupert Murdoch strode to a podium on a soundstage at the Fox movie lot in Hollywood on February 24, 1997. To his left was a large blue video map of the world. The name of Murdoch's global media empire, News Corp., was spelled out vertically on each of the four pillars on stage.

Wearing a dark, double-breasted suit, the tycoon saluted with his left hand, shielding his eyeglasses from the glare of red, blue and green spotlights that beamed down. He gained a better look at his audience: row upon row of industry analysts sitting at long, white-draped tables. The analysts, from firms such as Smith Barney and Goldman Sachs, had heard the expected reports that day on News Corp.'s financial health. They were told to stay late that afternoon for a special announcement, described by Murdoch as "our most ambitious project of all."

The pirate was about to swing his sword at the cable gang.

Reading verbatim from a press release, and thereby referring to himself in the third person, Murdoch announced that News Corp.'s American Sky Broadcasting would merge with EchoStar Communications Corp. in a $1.7 billion deal. Murdoch's Aussie accent elongated the *e* and softened the *r* of EchoStar to "Eeekostah."

The new alliance would simply be named Sky, matching several Murdoch satellite services around the world in an attempt to make the brand as well-known as Coca-Cola.

The partners would launch seven satellites into orbit to deliver several hundred TV channels, Internet and retail services down to the American masses. Subscribers would need only an 18-inch satellite dish that Murdoch later proposed selling for $50. The intent was clear: Murdoch would rain hell from above on TCI, Time Warner, Comcast, Cox and every other cable company. It was a prospect the cable gang had long feared and even given a talismanic name: Deathstar.

Like a James Bond villain—courtly in manner, global in ambition and narrow of intent—Murdoch looked the part to deploy a killer TV beam. Over and over, Murdoch had confounded the media establishment to build publishing, programming and satellite-TV businesses that checkered the globe.

After graduating from Oxford University, where he flirted with socialism and kept a bust of Lenin in his room, Murdoch became a staunch conservative. More accurately, he was a political opportunist who built his company on lowbrow entertainment while pandering to power. In Britain, his allegiance to prime ministers ranged from the Tory, Margaret Thatcher, to Labour's Tony Blair.

Murdoch viewed the 1974 resignation of President Nixon from the Watergate scandals as a crucifixion by the liberal U.S. news media. But he was a seasoned lobbyist who could work both sides of the aisle. Under Murdoch's direction, the *New York Post* hyped Republican candidate Ronald Reagan in the 1980 presidential election, but also endorsed Democratic president Jimmy Carter in that year's New York primary race against Senator Ted Kennedy. Murdoch wanted, and got, the Carter administration's support for a $7 million U.S. Import/Export Bank loan for Ansett, an airline Murdoch owned in Australia.

Murdoch's murky motives surfaced again in 1994. A News Corp. subsidiary contributed $200,000 to Republican congressional campaigns prior to the 1994 elections. After the GOP swept into Congress brandishing the Contract with America, News Corp.'s Harper-Collins inked a $4 million contract with Newt Gingrich for the House Speaker to write two books. Murdoch, who claimed to be unaware of the book deal, was at the same time seeking a waiver of Federal Communications Commission rules on foreign ownership of his Fox TV stations. Murdoch got the waiver. Gingrich gave back the book advance after a critical outcry.

Murdoch operated at a level at which politics, power, money and media were like a four-headed serpent. The Deathstar deal advanced his global satellite TV ambitions while burnishing his reputation as a mercenary.

His first love was news. It was the name of his company and reflected the black ink running in his veins. He inherited the struggling Adelaide *News* in Australia from his father, Sir Keith Murdoch, in the 1950s, then built it into the market leader. He bought the *News of the World* and *Sun* tabloids in Britain in the late 1960s, earning the nickname "The Dirty Digger" for his papers' diet of scandal, the occult and topless women. The latter innovation in daily journalism came when a *Sun* news photographer on holiday in St. Tropez snapped a bare-bosomed bathing beauty. The picture ran on page three of the *Sun*. The paper sold out, one editor noted, as if it were "printed on gold leaf." When puritans objected, circulation drove higher. Murdoch proclaimed, "I answer to no one but the public." His bid for respectability came later with the purchase of *The Times* and *The Sunday Times* of London.

Murdoch splashed across the Atlantic in 1974 with the launch of the *National Star* tabloid. Early features included "Killer Bees Head North" and a story claiming that "if all the Chinese jumped up and down in unison, the vibrations would cause a tidal wave that could engulf America." Once again using the edge of sensationalism to slip into the mainstream, Murdoch bought the *San Antonio Express* and *San Antonio News*, the *Chicago Sun-Times*, the *Village Voice*, the *Boston Herald*, the *New York Post* and *New York* magazine. But he would divest them all as he set sail for the electronic seas, primarily to comply with media cross-ownership restrictions.

Creating a fourth U.S. broadcast network to compete with ABC, NBC and CBS was considered folly. Yet, Murdoch launched Fox TV in 1986 after spending $1.6 billion to buy six major market TV stations from John Kluge's Metromedia and $600 million to buy the Twentieth Century Fox movie and TV studios from Denver oilman Marvin Davis. With the Fox purchase came studio chief Barry Diller, who fashioned Fox as a video version of Murdoch's tabloid sensibility. Shows like *Married ... with Children, A Current Affair, The Simpsons* and *America's Most Wanted* pulled in the 18- to 49-year-old eyeballs that advertisers covet.

Murdoch typically leveraged his way into partnerships, took over, then moved on to the next conquest. News Corp. did not dominate any particular sector. But like a drop of blood in a glass of water, the nature and notoriety of Murdoch's deals permeated and colored the entire entertainment world. "He's like a great white shark," John Malone once said in admiration. "He gets beached every once in a while, but he comes back and feeds."

With Deathstar, Murdoch headed, jaws agape, straight for Malone. This seemed odd at first. Murdoch had made his first foray into the clubby confines of cable programming in partnership with Malone, who held the keys to the kingdom.

"We're taking a friendly approach," Murdoch said in 1992.

TCI bought preferred stock in News Corp. and gave Fox affiliates cable channel positions low on the dial, which are typically viewed more often. TCI also carried Fox programming into more than a million cable homes in markets where Fox had no local broadcast stations. In 1993, Malone agreed to pay 25 cents per subscriber per month—$30 million a year—for TCI to carry Murdoch's FX channel, an amalgam of reruns and talk shows described by one critic as "the ultimate example of disposable television."

This made Ted Turner very unhappy. Not because of the quality of programming—Turner had deployed his share of reruns over the years from his Atlanta Superstation. He objected to the quantity of money flowing to Murdoch, whom he despised. "If the industry wants to spend a quarter for additional programming, why doesn't it spend it on me?" Turner whined in 1993, two years before his company's merger with Time Warner. "John doesn't seem to have a problem with multiple relationships."

Murdoch and Malone were just getting started. One goal was to create regional sports networks under the Fox Sports banner to compete with ESPN. That would be an expensive market-by-market smash mouth game, but Murdoch had proven himself cunning in the business of sports. He paid $1.6 billion for National Football Conference rights in 1994, intercepting CBS—the top network at the time—and giving Fox a marquee attraction. Murdoch wildly overpaid, some said, but he proved them wrong. Twelve broadcast affiliates, including eight from CBS, followed football to Fox later that year.

A newcomer to the American broadcasting scene, Murdoch moved boldly into cable programming. FX was the first entry, with Fox Sports, Fox News and Fox Family following throughout the 1990s. Yet his attempt to launch the conservative-slanted Fox News to compete with Turner's CNN in the early 1990s went nowhere. "I would have liked to start a news channel, but Malone and [Time Warner chairman] Gerald Levin would not give me the time of day," Murdoch griped in early 1994.

Malone shared Murdoch's ultraconservative political views. Both became board members of the Cato Institute, a libertarian think tank that would like to cut the federal government off at the knees. Both men were fans of Pat Robertson. Murdoch supported his presidential bid in 1988 and Malone helped Robertson restructure the finances of his cable network.

It wasn't a prerequisite for joining Cato, but Murdoch and Malone earned their stripes by avoiding corporate income taxes. In TCI's case, the net losses piled up and surged forward into future years like a red tide, erasing future income tax obligations. Murdoch took the international approach, setting up tax shelters and shell companies in the Cayman Islands and elsewhere to escape the taxman's bite in the United States and England.

So, Malone and Murdoch were on the same page when it came to political philosophy and tax strategy. But business is business, and Malone bit back after Murdoch publicly criticized him for not launching Murdoch's proposed cable news channel.

"Right now Rupert Murdoch gets 25 cents a month a subscriber for FX—Okay?" Malone told an interviewer later in 1994. "That's what he gets—a huge amount of money. And I'm wondering why in the hell I did that because he's running around saying, 'Well, they wouldn't do it for news.' I thought there was a low probability he could make a news channel work, given everything that's on news. But if he wants to pay me for access, I'd be happy to sell him access. You can tell him that."

Murdoch got the message. To get Fox News into 10 million TCI homes in late 1996, he gave TCI a $200 million loan and an option to buy 20 percent of the network. To make room, Malone cleared out existing networks like a bowling ball cracking into the tenpin. The arrival of Fox News in Denver pushed Court TV to split the programming day with Spice, a pay-per-view sex network. Viewers weren't sure if they'd witness heavy breathing in the courtroom or the bedroom. Murdoch typically paid other cable and satellite TV companies $10 per subscriber fee to carry Fox News, a channel that ended up in 38 million homes two years after launch.

The pay-to-play money was on top of $100 million to create the network itself. To head up Fox News, Murdoch picked Roger Ailes, the rotund Republican media adviser, onetime producer of Rush Limbaugh's TV show and former chief of CNBC, the Consumer News and Business Channel. "Fox News is going to be balanced and fair and that might look a little more conservative to people who've been tipping to the left," Ailes said. "I do notice that it is terrifying some reporters."

It certainly terrified Ted Turner.

Murdoch reportedly offered Time Warner $25 a head to launch Fox News, but Turner would have rather yanked out his own silvery half-mustache hair by hair. Turner is an unabashed liberal with many passions, not the least of which is his carnival hatred of Murdoch. Turner once challenged Murdoch to a pay-per-view boxing match, which Murdoch dodged. Turner claimed to dream of Murdoch's death and compared him to Adolf Hitler. Murdoch, according to Malone, said he didn't mind the comparison.

Turner kept CNN afloat through a sea of doubters and financial losses in the 1980s. The network earned acclaim and profitability through wall-to-wall coverage of the fall of the Berlin Wall, the Gulf War and other hot spots. But CNN was vulnerable to competition and got protection from the cable gang. In the late 1980s, NBC proposed launching a cable news channel as a competitor to CNN. But TCI and Time Warner, part-owners of CNN, would not carry it. Blocked, NBC stuck with its CNBC format on a channel it bought from TCI for $20 million. Tennessee Senator Al Gore described this payment by NBC as a "shakedown" by TCI.

Gore was a self-appointed prosecutor of Malone and the cable gang. He gave voice to the very real concerns about their market stranglehold and skyrocketing cable rates in his home state and elsewhere in the late 1980s. Yet, Gore was also a partisan of satellite TV who trashed the cable gang for his own political gain. When offered mitigating evidence or opinion, he dismissed it, preferring to cast their actions in the darkest possible light, à la the "Cable Cosa Nostra." The cable gang's failure to squelch rate abuses within their industry was a huge tactical error. It allowed Gore and others to define the terms of the debate and ended up costing them billions of dollars from federal rate caps in the 1992 Cable Act, shortly before Gore became vice president under Bill Clinton.

Malone was willfully oblivious to government relations, declining to bow down before politicians whom he viewed as hopelessly irrational and hypocritical. Gore critiqued a "shakedown" by TCI. Yet a $100,000 donor to the Democratic National Committee would later describe a Gore phone solicitation from the White House with the same word: "shakedown."

Murdoch was able to launch Fox News onto systems everywhere but in its hometown of Manhattan, where Time Warner and Ted Turner wielded the cable pipe. However, as part of the government's antitrust approval of Turner's merger with Time Warner, the company was obligated to carry a 24-hour news competitor to CNN. Time Warner chose MSNBC, the cable news venture of Microsoft and NBC, bankrolled with $220 million by Microsoft chairman Bill Gates. Time Warner could have carried both MSNBC and Fox News, but Turner vetoed that idea. Murdoch was denied one million potential viewers in New York City alone. The only way to see Fox News in the world's media and advertising capital was to walk by its sidewalk studios at News Corp. headquarters off Sixth Avenue.

Murdoch was furious and Ailes pumped him up. "If these guys want to go to war," said Ailes, "let's go to war."

Murdoch had bought back the *New York Post* out of bankruptcy in 1993 and sometimes used it as his personal printing press. The *Post* questioned Turner's sanity by caricaturing him in a straitjacket. It also dropped CNN listings from its paper. During a World Series game between the New York Yankees and Turner's Atlanta Braves, Murdoch hired a plane to fly over Yankee Stadium with this blinking message: "HEY ... TED ... BE ... BRAVE ... DON'T ... CENSOR ... FOX ... NEWS ... CHANNEL."

Murdoch filed an antitrust lawsuit against Time Warner and enlisted Republican muscle, including New York mayor Rudolph Giuliani, who proposed using one of five city-controlled public access channels on Time Warner's cable system to carry Fox News. Malone, whose company owned 9 percent of Time Warner and was a partner in myriad ventures with News Corp., was tickled by the uproar.

"Rupert's worth $3.5 billion. Ted is worth $2 billion," Malone said. "Time Warner is the largest media company in the world. Rupert's got to be a close second. This ought to be regarded by everybody as free enterprise jousting. Guys duking it out, trying to gain the public's attention, trying to use the system to leverage each other. I would waste no tears on either one of these guys. In the end it will get resolved by some transfer of economics."

Though no one could know it then, the transfer of economics would soon involve Deathstar and Malone's own considerable leverage on behalf of the cable gang.

MURDOCH, THE UNDISPUTED king of satellite TV worldwide, was set on adding the final jewel to his crown: the U.S. market. The cost of launching Fox News on cable and the difficulty dealing with Time Warner only heightened his desire to own satellite distribution.

Like Gates, Malone, Turner and other power brokers in the interlocked communications world, Murdoch understood and trafficked in just one currency beyond money: leverage. With it, you get your way. Without it, someone else gets your way. Competition is the most radical form of leverage. If Time Warner or any other cable operator declined to carry Fox programming or required up-front payments or gave lousy channel positions, what would they do when Murdoch lured cable customers with $50 rooftop satellite dishes that offered double the number of cable channels?

There are many ways to play cutthroat. Plus, Murdoch already had buckets of money invested. A year earlier, Murdoch, Malone and Charlie Ergen took part in a wild bidding battle for the last full satellite TV slot in the United States. Murdoch and his partner, MCI Communications Corp., paid the staggering sum of $682.5 million at federal auction for the slot, the galaxy's most expensive parking place. "I think we're moving to an increasingly wireless world," Murdoch explained. Others had figured that out for a lot less money. The remaining two full satellite slots were controlled, one each, by DirecTV and Ergen's EchoStar. As pioneers of the fledgling small-dish satellite TV business, they received the slots virtually free of charge from the federal government.

As with the creation of Fox TV and the payment of big bucks for NFL rights, Murdoch's sanity was questioned after the auction. He and MCI paid more than two-thirds of $1 billion for air. They would need to pump in perhaps $2 billion more to get satellites launched, dishes manufactured and the service marketed—all for the prospect of being late in the game. The News Corp./MCI venture was called American Sky Broadcasting, ASkyB for short. They began building a $100 million satellite uplink facility in Arizona and solicited pitches on Madison Avenue for a $75 million advertising account.

All the while, to mitigate the risk, Murdoch separately negotiated with the men he had beaten at auction, Malone and Ergen.

A joint venture with Malone was complicated by the tangled ownership of the cable gang's Primestar satellite TV service, part-owned by Time Warner.

The alternative was Ergen.

He was 22 years Murdoch's junior, but cut from the same rough cloth, with a passion for satellite technology, a stomach for billion-dollar debt and a flair for the dramatic. After the 1996 satellite slot auction, Ergen flew out to Los Angeles for a meeting with Murdoch and Michael Milken. The former junk bond king had served 20 months of a 10-year prison sentence for securities fraud, ending in 1992, after prosecution by Rudolph Giuliani, then an ambitious U.S. attorney.

Milken had been mother's milk for upstarts such as MCI, News Corp., Turner Broadcasting, TCI, McCaw Cellular and Viacom. Working from an X-shaped trading desk at an outpost of Drexel Burnham Lambert in Los Angeles, far from the conservative bank-ing establishment in New York, he threw the dice with the entre-preneurs, funding their fledgling communications companies with high-interest junk debt in the 1970s and 1980s. The milk turned rancid with get-rich-quick leveraged buyouts and characters like inside trader Ivan Boesky, who won a plea bargain with Giuliani by wearing a wire to take Milken down. Boesky's contribution to American culture was immortalized in the 1987 movie *Wall Street*, with Gordon Gekko's line, "Greed is good."

Boesky actually said, "Greed is all right," which sounds all wrong.

Though barred from the securities industry for life, Milken re-mained a financial strategist and shadow adviser to his old media pals. Turner paid Milken a $50 million fee for a week's work in the Time Warner/Turner Broadcasting merger. Milken also brokered a broad alliance between MCI and News Corp. He was a big believer in satellite technology and urged them to acquire the last available U.S. satellite slot at auction, which they did. Milken was even higher on a merger with EchoStar and its potential to rule the skies. "This deal will change the world," he told Murdoch and Ergen during their meeting.

Ergen was a relative nobody in the entertainment and technology industries. Yet, here he was plotting to change the world with Rupert Murdoch and Michael Milken, the man who popularized the very financial instrument—the junk bond—which Ergen used to fund his own company's attack on the cable gang. Their conversation did not result in a deal then, but within a year, Murdoch and Ergen would try again on their own.

Ergen had no leg up and no early patron as did Murdoch and Malone. But he was whip smart and hungry to play in the big leagues. He described himself as "a country boy from Tennessee" just selling satellite dishes. But the aw-shucks routine masked a nut-cutting competitor who would slice prices to the bone to gain market share. "A dish in every home!" Ergen proclaimed at industry events. "Cut the cable and throw it in the trash!" Trim and boyish in his mid-40s, Ergen bypassed the suit-and-tie uniform of corporate America in favor of a button-down shirt with the company logo over the left breast. Khaki pants and a preppie cloth belt with a sailboat pattern completed the ensemble. He positioned himself as an outsider, a crusader saving America from cable rate hikes and shoddy service. After succeeding in the big-dish satellite market in the 1980s, he acquired a small-dish satellite license from the government, bought out other companies and ended up with more orbital real estate than any other player in the United States.

While Ergen had the nerve, or naiveté, to take on the cable gang, he lacked the track record and leverage of kingpins like Murdoch and Malone. EchoStar's 150-channel service, named Dish Network, was bounced from an uplink center in Cheyenne, Wyoming, to two satellites 22,300 miles above Earth. The video beams rained down on pizza-sized dish receivers bolted to homes and aimed toward the southern sky. EchoStar gained 500,000 subscribers in 1996, its first year of operation. It was a healthy, though relatively minuscule, number. At TCI, Malone bought cable systems that size before driving home for lunch with his wife.

Given the technological and competitive waves that smash through the consumer electronics business, capsizing the little guys first, there were doubts that EchoStar could survive without a well-funded partner. Carl Vogel, EchoStar's vice president, told Wall Street that EchoStar would have one by year-end 1996. A potential partner was Vogel's previous employer, Jones Intercable. Other names that surfaced were Sprint Corp., Lockheed Martin Corp. and Paul Allen, the billionaire investor and Microsoft co-founder who owned 4 percent of United States Satellite Broadcasting, a satellite TV service in partnership with DirecTV. None of those deals happened and EchoStar's stock price fell by more than half, to $18, by the end of January 1997.

Murdoch rode in like a white knight. And if Ergen did not accept Murdoch's hand in partnership, he would soon face it raised as a competitor.

On February 7, 1997, Murdoch's corporate plane flew into Centennial Airport, a small airfield south of Denver. Limousines sometimes stand by to ferry the jet set to their destinations on the ground. Judianne Atencio, EchoStar's spokesperson, was waiting in the company's Chevrolet Suburban to take Murdoch back to company headquarters for a meeting with Ergen.

At one point in their ongoing negotiations, Murdoch told Ergen, "Those cable guys are greedy. They don't want to put my programming on. I'll show them. I'll bring them to their knees." Ergen ended discussions with Sprint, while EchoStar went into stealth mode. Executives locked their offices when they walked down the halls and shredded papers after meetings. Murdoch flew back again the next week, on Valentine's Day.

"Chahlee," Murdoch purred to Ergen one night at dinner. "You remind me of myself when I was younger." Ergen fell under the older man's considerable charm, as had a string of business associates seduced by Murdoch over the years. Muhammad Ali had the Rope-a-dope. Murdoch had the Rupe-a-dupe. "He has this fatal capacity to instill the confidence in you that you and he have a special, exclusive relationship," said Harold Evans, a former editor at Murdoch's *Times* of London. "It's a wonderful con trick."

Murdoch visited the Ergens at home. The couple's five children—none of them yet a teenager—surrounded the man whose Fox movie studio made the smash hit *Independence Day*. The Ergen kids were not allowed to see such PG-13 movies. They asked Murdoch why he didn't make action movies they could see.

"You know what," said Murdoch, charming and avuncular. "You're probably right."

Murdoch and Ergen moved closer to a merger that looked like a winner on paper. Ergen had two satellites in orbit, a low-cost strategy and a proven dealer network. Murdoch had key programming, marketing savvy and money. He would contribute the 110-degree satellite slot, the Arizona uplink center, two satellites and cash. The deal was valued at $1.7 billion, which included the money paid at auction for the slot. Once they came to terms, Murdoch wanted to move fast to announce it at News Corp.'s analyst and investor meeting. He was facing hard questions about his satellite TV plans and the cash drain they represented. A tabloid journalist at heart, Murdoch also had the timing and instinct for grabbing headlines.

On Thursday, February 20, Ergen and Murdoch signed a letter of intent outlining their partnership, which gave them an equal number of seats on the board of directors. Yet, Ergen gained just a hair more control of the company's voting power and locked in provisions that would make it tough for Murdoch to back out. It seemed out of balance, considering that News Corp.'s revenues in a week equaled EchoStar's in a year. "He can't fire me and I can't fire him," Ergen said later. "It's like a marriage."

Atencio and other EchoStar officials flew out to Los Angeles, where News Corp. had a suite atop the Century Park Hotel. The room was filled with electronic gear to create visuals for Monday's merger announcement. "We were used to doing things on an Etch-a-Sketch," said Atencio. "We were like the country cousins."

Ergen, Vogel, Atencio and others went down to the hotel bar for a drink on Sunday afternoon, giddy with anticipation. They pulled out their EchoStar business cards and wrote in titles to reflect their new positions with Sky. Ergen crossed out "chairman" and wrote "chief executive officer." Vogel put "executive" in front of "vice president." There were some hard feelings about this, since Vogel expected to be named president of the merged company. Murdoch had approved it. Ergen had not.

Murdoch would become chairman of Sky. Preston Padden, his newly named chief of worldwide satellite operations, was to keep an eye on Ergen. Other ASkyB officials would fill various executive positions in the merged company. The deal with EchoStar put Murdoch in the enviable, though complicated, position of supplying Fox networks to the cable gang on one hand and threatening to steal their customers by satellite on the other. But Murdoch, as he had done so often in the past, would shoot first and notify the next of kin later. Prior to his appearance on the Fox soundstage to address the analysts, Murdoch held a hasty press conference. In the room were reporters from the *Los Angeles Times, Business Week* and *Fortune* magazine, among others. Calling in by phone were reporters from papers including *The Wall Street Journal, The New York Times* and *USA Today.*

"We're aiming for the big cable market, 65 million homes," Murdoch told them. "We expect to have a good 50 percent of all new satellite customers from here on."

"Why don't you want the other half?" one reporter asked, prompting laughter because everyone knew that Murdoch always wanted both halves of everything.

"There will be a maaaja mahketing launch," said Murdoch, stretching out the *a*'s and dropping the *r*'s. "Maaaja."

The prospect of Murdoch throwing billions of dollars into the Sky merger with EchoStar was tantalizing. Cable was already struggling with DirecTV, owned and funded by mighty General Motors. During a historic bull market, cable stocks were in the tank. Now they faced Murdoch and Ergen joining forces to lure even more subscribers away, perhaps never to return.

On stage, Murdoch ended his brief statement by saying, "I'd now like to invite my good friend and new partner, Charles Ergen, together with Preston Padden, to join me on stage and tell you all about what they're going to do to take over, at least, North America."

These were two men with a deep and abiding antagonism toward the cable industry.

Ergen had taken TCI and the cable gang head-on in the summer of 1996, dropping the price of a satellite dish from $499 to $199, targeting cities where cable rates had gone up 20 percent in some cases. The reaction at TCI to Ergen's kamikaze attack was bewilderment.

"He's crazy," Malone told associates. "He doesn't want to make any money."

Ergen ate it up. "Tell 'em I'm the crazy guy!" he said. "Crazy Charlie!"

EchoStar attacked cable at its most vulnerable points: rate hikes, limited channel choices and a lack of sharp digital pictures. Ergen dismissed the cable industry as monopolists squeezing a profit from rotting analog wire in the ground when the future was clearly in the sky. Unlike the rest of the cable gang and its programmers, most of whom were obliged to kiss Malone's ring, and sometimes lower, Ergen had no problem speaking his truth to power.

"The problem is, he's got $15 billion in infrastructure," Ergen once said, referring to Malone and to TCI's embedded cable investment. "He's in the same place General Motors was when Japanese carmakers came in. They didn't want to write their tooling off."

Now here was Ergen, center stage in Hollywood, with the chance to tell the world how he was going to shove it down the cable gang's throat. Yet, the first words out of his mouth reflected the haste with which his deal with Murdoch had been struck. "Rupert didn't have it exactly right," said Ergen. "He didn't tell you everything—the truth. He knew full well two weeks ago he was doing this deal. The only problem was, he didn't tell me until this weekend that he was doing the deal."

Then his slide show was delayed. Ergen had dead air to fill. He looked out on the sea of expectant faces, including industry analysts who had not paid much attention to satellite TV stocks in general, or EchoStar in particular. He lit upon John Reidy, a Smith Barney analyst. Ergen proceeded to tell a joke, the punch line being that Reidy's brain would command a higher price than others because it had never been used.

Mercifully, the slide show clicked into place and Ergen began his pitch. One slide showed the word "Cable" in yellow, enclosed in a red circle with a diagonal slash through it. "We believe the world is going wireless and you don't need cable in your house," said Ergen, claiming that Sky would put the equivalent of the "fiber optic superhighway on the rooftop of every square inch of the United States in the next 18 months."

Ergen's brief presentation was overshadowed by Preston Padden, the blustery chairman and chief executive of ASkyB. Padden was once a lobbyist for independent TV broadcasters. In the late 1980s, he eagerly testified in front of congressional subcommittees against Malone, who claimed that subscribers "vote" for cable every month by paying their bills. "Given the monopoly character of cable, these 'votes' are equivalent to the 'votes' received by the Communist Party in pre-Perestroika Russia," Padden replied.

On the Fox soundstage, Padden hyped Sky and slammed cable in a script he had practiced in the hotel suite the previous two days. "The easiest way to visualize what we'll be doing is to imagine a highly cost-effective and functionally superior wireless overbuild of the entire American cable industry," said Padden. "Sky will finally deliver on the original cable prophecy of a 500-channel universe." These remarks contained two big digs at the cable gang. "Overbuild" is to cable what "wooden stake" is to Dracula. It refers to building a competitive wired or wireless network to deliver subscription TV service to homes and businesses. The 500-channel comment was aimed at Malone, who in 1992 publicly fantasized that the cable industry would offer that many channels but had never delivered.

"Our goal is to come to market with a television product so superior, and a consumer proposition so compelling, that a substantial number of 70 million households stop writing their checks to their current service—usually cable—and start writing them to Sky," Padden said. Grinding his boot heel a little deeper into Malone's back, Padden highlighted the cable industry's massive debt, eroding subscriber base and huge costs to rebuild for new services. He then outlined Sky's ambitious plans: four satellites blasted into orbit in the next 18 months, joining two of EchoStar's already in place, plus one more down the road, giving Sky "a cosmic armada unmatched since the Empire Struck Back."

This was an allusion to the original *Star Wars* movie trilogy that Murdoch's Fox Studios was re-releasing into theaters. And, if one wanted to fish for deeper symbolism, it referred to vanquishing Malone, whom Gore had once knighted "Darth Vader."

Sky would prevail, said Padden, by redressing satellite TV's biggest failing: local channels. Technical and regulatory restrictions prevented satellite TV from offering popular broadcast signals to the vast majority of subscribers. Therefore, no local news, sports and weather—a big disadvantage compared to cable. For all of its rate hikes and service problems, cable was a local business gone national. Satellite TV was national and needed to go local.

Padden said Sky would do just that, delivering local channels to 75 percent of U.S. households within a year, plus all the existing cable programming and 60 pay-per-view channels. Sky would provide subscribers with "eight channels of HBO for the same cost as the single channel they get from their chintzy and capacity-constrained cable company," Padden said. Sky would align with MCI, and very possibly with the Baby Bell phone companies, to distribute the service. Padden's kill shot was a line that would define and haunt the Sky deal for months to come. It was repeated in news accounts and conversations as shorthand for all that was at stake. With Murdoch and Ergen sitting nearby, Padden predicted that Sky's assault on the cable industry would be so harsh and unrelenting—yet so attractive to consumers—that Malone and the cable gang would commit a business version of euthanasia.

"At that point," said Padden, "the cable guys will be calling for Dr. Kevorkian."

—

GORDON CRAWFORD COULD not believe his ears. It was an outrage. And the senior vice president of Capital Research and Management Co. was not easily outraged. Capital Research is a mutual fund company based in Los Angeles that controls billions of dollars in assets. Crawford handled the entertainment and media investments, managing hefty stakes in TCI, Time Warner, News Corp., Viacom, Disney, America Online, Comcast, CBS and many others. In some cases, Capital was the single biggest investor, which gave Crawford's opinion great sway. "Gordy," as he is known to his friends, was thoughtful, bespectacled, a churchgoing family man and by no means a household name. Yet, he was as big a player in his own way as Murdoch, Malone and Turner. With them, he was annually cited on the list of *Vanity Fair*'s 50 top Information Age power brokers. Crawford's role was that of consigliere, an adviser. His mission was not to build an empire for himself, but to make sure that the empire builders and their oversized egos did not ruin his multibillion-dollar investment portfolio.

What Padden proposed sounded like ruination.

"The presentation of the deal was the most egregiously stupid presentation by a corporate executive that I've ever seen," said Crawford, still hot about it two years later. "I went up to Rupert afterward and told him that talk was going to cost him a lot. I don't think he realized that the cable industry is this small little club of guys whose net worth is tied up in their companies and how pissed they would be."

Murdoch laughed and agreed that Padden went overboard. Yet, Crawford was one of the doubters in 1989, saying that Murdoch "grossly overpaid" for the Metromedia stations. Murdoch proved them all wrong. He always had and he always would.

Murdoch undoubtedly played the media coverage like a master. The hype gushed across the country. "Building a TV Titan in the

Sky," headlined *USA Today*. "Deal by Murdoch for Satellite TV Startles Industry: An Alternative to Cable," declared *The New York Times* in a front-page story. "Murdoch, Ergen Dish Cable a Blow" in *The Denver Post*. "Is cable television dead?" began a *Bloomberg News* report.

EchoStar's stock shot up $3 to $18 before trading was halted on the day the deal was announced, then up another $8.75 the next day.

Cable stocks, already troubled, fluttered south. The cover of *Cablevision*, an industry trade magazine, portrayed acid raindrops containing Murdoch's face, with the headline "Rain of Terror?" An inside story was headlined, "Cable to Murdoch: Drop Dead." That expressed the gut reaction of the cable gang meeting at the industry's national convention in New Orleans three weeks later.

"We're going to make it as tough for [Murdoch] as we possibly can, kind of like the Russian army did with the German army," said Turner, leaving listeners to understand that if Murdoch was Hitler, Turner was Stalin.

Beyond staging legal roadblocks in Washington, the cable gang threatened to block expansion of Murdoch's Fox sports, news and entertainment channels onto their cable systems.

"I don't think, generally, people want to buy bullets for the gun that's going to shoot them," said Glenn Jones, whose Denver-based cable company served more than one million subscribers. Marcus Cable in Dallas refused to even meet with Fox sales executives pitching Murdoch's FX and Fox News channels. "If someone is threatening to burn your house down, you don't invite them in for lunch first," said Jeff Marcus, the company's chairman. "We are not going to give them money so they can build a competing satellite business."

And then there was Malone.

He had previously discussed separate satellite deals with both Ergen and Murdoch, but could not come to terms. The three men were like brothers in a morality play. Each of them celebrated a birthday in early March and they were almost equally separated in age. In 1997, Murdoch turned 66, Malone, 56, and Ergen, 44. The older and younger warriors were joining forces against the middle one when he was most vulnerable.

Malone had seen the potential, and potential threat to cable, of satellite TV as dish sizes decreased, channel capacity increased and competition mounted over the years. "I remember having conversations in the mid-1980s where John and I would talk about it," said Trygve Myhren, a former cable company chief. "It was the one thing we were most afraid of."

Fear sharpened Malone's wits.

Within a week of the Deathstar announcement, Malone pressed ahead with a $250 million satellite launch from Cape Canaveral, Florida. It would go to a slot shared with an EchoStar satellite already in orbit. The 100 or so new channels provided might serve as an adjunct to basic cable in some areas, or be used by Primestar, stiff-arming DirecTV and EchoStar from gaining more subscribers at the cable gang's expense.

Malone flew to south Florida to witness the satellite launch. With other officials, he signed a photo poster of the rocket and satellite. The poster was hung in the mission control bunker near the launch site. Malone also gave an interview to *Satellite Business News*, an industry trade publication.

"I doubt that they will get along very well," said Malone, referring to Murdoch and Ergen. "Rupert is so aggressive that he really doesn't make a great partner."

This comment seemed to say more about Malone's situation than Ergen's.

The Deathstar deal was the firepower Ergen needed to compete head-on with the cable gang. From satellites one-tenth the distance to the moon, Murdoch and Ergen could cast commercial thunderbolts against cable and reap profits they estimated at up to $1 billion a year. That threatened the cable industry power base that Malone had built deal by laborious deal over the previous 25 years.

With the satellite launch delayed by weather one day and by a technical problem the next, Malone left Cape Canaveral before the

Loral satellite, atop a Lockheed Martin rocket, blasted off the launchpad. And when it did, ripping a furrow in the blue sky before sending the satellite into geosynchronous orbit, there was a problem. The satellite had an internal power glitch that would degrade its expected 12-year life and crimp its ability to deliver the expected 100 channels. It was no Deathstar.

In early 1997, the business they had dominated for decades was not going well for Malone and the cable gang. The momentum of the game had suddenly shifted, which raised a provocative question: Given cable's 50-year history of technical innovation, 65 million subscribers, annual revenues of $25 billion and a local monopoly in almost every American town, how had it all gone so wrong so fast?

THREE

STEALING
FREE TV

R OY BLISS PILOTED the four-seater Cessna through the
Wyoming sky high above the Big Horn Basin, a valley
ringed with mountains and named after the bighorn sheep that
roamed the rocky ledges. The plane's passenger door was cracked
open a few inches, letting the wind whistle in at 8,000 feet.

John Huff sat in the passenger seat with a radio frequency meter
in his lap. A wire ran from the meter to an antenna tightly taped to
the outer strut of the high-wing aircraft.

Bliss and Huff were hunting for free TV.

They lived in Worland, a town located midway between Casper
and Yellowstone National Park in northwest Wyoming. One question captivated Worland's 7,000 residents in that summer of 1952:
What size antenna tower could pull in the lone television signal
broadcast from Billings, Montana, 125 air miles to the north?

Bliss, a former air force pilot, owned the Culligan soft water franchise in Worland. He flew the Cessna for fun. Huff, who managed
two-way radio operations for the local Pure Oil Co., knew antennas. They flew west of Worland at a cruising speed of 125 miles per
hour. The dipole antenna was like a fishing pole, with Huff moni-
toring the meter.

"We didn't even get a whisper," Bliss recalled in an interview 46 years later. "We were pretty disgusted."

Bliss banked the airplane to a lower elevation and cruised by a new wildcat oil well in the Big Horn Basin. He dropped lower and lower, no more than 25 feet off the ground, when the meter needle jumped in Huff's lap.

"Hey!" he said. "We've got a lot of signal here."

They eyed the spot, which was on federal land, flew back to the airstrip, then drove back to confirm the meter reading. Within a few weeks, they had an antenna, a television set and a gasoline-powered engine to generate electricity up at the site. When it was all hooked up, Bliss pulled the TV set knob and clicked the dial to Channel 2. Up popped a black-and-white picture from KOOK in Billings, an independent TV station unaffiliated with ABC, NBC or CBS. Its broadcast day, from 4 P.M. to 11 P.M., featured local news, *Howdy Doody* and *Boston Blackie*, a detective show.

"We knew we were on to something," said Bliss.

That they captured a TV signal just three feet off the ground was a fluke. The KOOK signal beamed from a Billings transmission tower dimmed considerably beyond a 50-mile radius. But when it hit the Pryor Mountains outside Worland, it deflected down into the valley, forming a cone of signal detected by Huff's antenna gear. Bliss and Tom Mitchell, an electrical contractor and former army radar technician, were intrigued by the possibility of carrying the TV signal four miles into Worland. Bliss sold his Culligan soft water franchise and used the money to help start Western Television Corp. with Mitchell. Huff, the man with the antenna, stayed at his oil company job.

"It was a shaky deal," said Bliss, then 34 years old and married with two children. "We didn't know what we were doing. Nobody knew what cable was. We couldn't even find equipment."

Bliss and Mitchell cemented a wooden pole into the ground at the signal site and attached a basic rooftop antenna to it. Mitchell bought a black-and-white DuMont brand TV set, housed in a wooden cabinet, for $300 in Denver. They had a contractor build an

8-by-10-foot brick building and haul it up to the site on a flatbed truck. Into the building went the TV set, the generator, a wood stove and an old sofa that seated four. Bliss draped a black curtain in front of one wall and cut a hole in it to frame the 17-inch-diagonal screen of the TV set. That winter, when the temperature dropped far below freezing, Worland townsfolk drove the dirt road out of town to see TV—some of them for the first time. A few of the curiosity seekers, more familiar with movie theaters, thought there was a projector in the back of the TV set. The mayor and members of the Worland town council came out for a look. Others crowded around the building's entrance or sat in their cars, waiting their turn, the Wyoming night black save for the headlights, the pinpoint stars above and the black-and-white glow of a TV that was like a modern discovery of fire.

Bliss and Mitchell flew to Walla Walla, Washington, to learn about a cable TV system there. They began buying coaxial cable and amplifier boxes. From their makeshift antenna, they laid four miles of cable into Worland, most of it underground. They crossed the Big Horn River by putting the line up on poles owned by the power and phone companies. Pole attachments would loom large in subsequent years for the cable gang. American Telephone and Telegraph, better known as AT&T, had its hands full after World War II, filling two million back orders for phone service. AT&T's Bell Labs had patents for coaxial cable, but the phone company wasn't aggressive enough in trying to deliver broadcast signals through it. AT&T did, however, come to see cable as a competitive wire to the home, which led to bruising legal and regulatory battles regarding pole attachment and other access issues.

But first, the early cable cowboys faced a daunting array of engineering and financial struggles.

One banker in Worland denied Bliss a loan for the project. Another agreed, but only if Bliss and Mitchell could get the TV signal into the town hall. Appliance dealers in Worland had TV sets in stock, but few were sold because there were no signals to receive. Mitchell convinced the dealers to display 30 sets at the town hall for a demonstration. He then drove Worland's mayor and the heads of the local phone and power companies to the KOOK studio in Billings. They discussed live on camera the possibility of delivering TV signals to homes by cable—an impromptu public affairs program seen by Worland locals packed into the town hall, and by viewers in Billings already receiving the TV signal through rooftop antennas. "Cable was kind of a hard sell because so few people had a TV anyway," Mitchell recalled in 1998. "You had to spend $300 to $600 for a television, which got you one channel for six or seven hours a day."

Then, there were the cable fees. Once it wired parts of Worland, Western Television billed subscribers $6.50 a month, plus an upfront installation charge of $150. The payment plan idea came from rural electric companies, which charged a similar connection fee to bring power to home owners.

"In one sense, the local operator was a hero," said Stratford Smith, a cable industry attorney beginning in the 1940s. "But life was hell because people didn't want to pay for television. They knew everybody else got it for free. Why should they have to pay?"

By the fall of 1953, a year after the airborne capture of the KOOK signal, Western Television had 100 subscribers in Worland, barely enough to stay in business. The system required amplifier boxes half as big as a refrigerator, placed every 2,000 feet, to deliver a clear signal through 30 miles of cable in town. Each amplifier box contained 12 tubes that occasionally burned out, causing subscribers' TV sets to drop sound, scatter pictures or display ghosts and snow on the screen. Roy Bliss Jr., eldest son of the company's founder, helped change the thumb-sized tubes, burning his fingers as he pulled them out of the narrow sockets and replaced them on the fly. Over the next two years, Western Television expanded into five nearby towns: Lander, Riverton, Greybull, Basin and Thermopolis, a onetime hangout of Butch Cassidy and the Hole in the Wall gang.

In the late 1940s and 1950s, small towns all over rural America got their first taste of broadcast television from entrepreneurs like Bliss and Mitchell. The service was known as Community Antenna

Television, or CATV, sometimes advertised as Snow Free TV. Paying for what those in the cities received free grated on some, but at least rural residents had access to TV signals previously blocked by geographical and technical barriers. "The theater, the world of good music, lectures, big-time sporting events and many other items cannot be afforded by a small population," Montana state senator Frank Hazelbaker wrote years later to Dillon Cable TV in his state. "Until the advent of cable TV, we in small places were isolated from many of the finer things in life. Now it is a different picture."

One of the nation's first cable hookups came courtesy of a radio station operator named Ed Parsons. His wife, Grace, nagged him to get TV into their home in Astoria, a coastal town in Oregon, 125 miles southwest of Seattle. After some investigation, Parsons rigged an antenna atop the John Jacob Astor Hotel in Astoria and ran a line to the TV set in his nearby penthouse. Inside, Grace used a walkie-talkie to report to her husband on the reception from KRSC-TV in Seattle. That was Thanksgiving Day 1948, at which point their living room became a tourist attraction. So, Ed Parsons dropped a cable line down to the hotel lobby and then across the street to Cliff Poole's music store, setting up other TVs for public viewing. Parsons later moved to Alaska and became a bush pilot.

In Mahanoy City, Pennsylvania, an appliance salesman named John Walson found that the mountains blocked his town from receiving broadcast signals from Philadelphia, 75 miles to the southeast. Walson mounted an antenna atop Tuscarora Mountain and strung a wire into town. Viewers tuned into shows like the *Kate Smith Hour, Pulitzer Prize Playhouse* and *Beat the Clock*.

TV had thrilled the United States at the 1939 World's Fair in New York. By 1952, there were 17 million black-and-white sets in use nationwide, but only 14,000 hooked up to cable. It was that year when Bill Daniels, age 32, stopped to wet his whistle at Murphy's Bar in Denver. There he saw TV for the first time, the Wednesday Night Fights sponsored by Pabst Blue Ribbon. Daniels, a former Golden Gloves boxing champ in New Mexico, fell hard.

"I thought, 'My God, what an invention is this!'" he recalled years later. "I couldn't get it out of my mind—how do you get that great invention to a small town that didn't have any TV stations?" Like many of the early cable guys, Daniels was a World War II veteran who brought a gung-ho spirit and battlefield communications experience into a new peacetime business. Daniels went on to build cable systems in Casper, Wyoming, and throughout the West. The hookup fee was $150 with a monthly charge of $7.50.

"Every 90 days we would send our customers a poll and they'd pick what programs they wanted to watch—it was just black and white [TV] then," Daniels recalled. "The majority ruled. If more people wanted to watch *I Love Lucy* than *Sid Caesar*, then that's what we showed." Gene Schneider, who worked with Daniels and became a cable titan in his own right, recalled it a little differently. "We sent out ballots on what to watch," said Schneider. "Then we'd pick what we'd like to see."

Like many of the cable cowboys in the West, Daniels settled in Denver because it was a Rocky Mountain crossroads. His success over the years provided him a palatial home he named Cableland, which looked like a Goldwater Republican version of a Playboy mansion. It had marble floors, bottles of Dom Perignon behind glass, a womb-like mauve interior, a sunken den with 64 tiny TVs, shelf upon shelf of GOP elephant sculptures and an outdoor pool in which floated a yellow toy duck. On an upstairs bedroom wall was a laminated plaque with Daniels's mini-biography in the 1993 *Forbes* magazine issue listing the 400 richest Americans. The staccato paragraph hits the rhythm of a midcentury businessman's roundabout rise:

Cable TV. Denver. 73. Four divorces, two stepchildren. Born Greeley, Colo.; to New Mexico at 14. Graduated N.M. Military Institute. WWII Navy fighter pilot in Pacific. Joined father's life insurance business in 1946, didn't like it. After Korea recall 1950 to 52, to Casper, Wyo.; started oil insurance business. Laid cable to bring TV to Wyo. Went to Denver 1958 to form Daniels & Associates, built to top cable brokerage. Also acquired cable systems. First sale, $100 million, to Newhouse 1980; second sale to United Artists 1989.

"They made offers that we couldn't refuse." Overcame severe alcohol problem. Today, brokerage, remaining systems, other interests estimated $300 million.

In poor health in later years, Daniels resided mostly in California. He loaned the $7 million Cableland mansion out for charity events and willed it to the city of Denver as a mayor's residence. "Take a look at where this lucky guy lives a most fortunate life," Daniels said of himself, announcing the handover to Denver mayor Wellington Webb in 1997.

Daniels was cable's most influential broker, the point man between buyers and sellers as the industry grew and then consolidated. He also played a key role in cable sports programming and even suggested that the Super Bowl might someday be seen on pay-per-view. Asked about the cable gang's reputation as tough guys, Daniels took a draw on his ever-present More cigarette and said, "We were just scratching. Your tough guys entered the business when Wall Street entered the picture."

In the 1950s, Daniels scouted and brokered properties for many new cable operators, including Bob Magness.

Magness was born an only child in Clinton, Oklahoma, in 1924 to a hardscrabble farm and railroad family. His father's father fell to his death after stepping off a cattle train parked on a bridge. His mother's father helped build western railroads, with the family living in tent cities along the way. Magness hunted and fished and helped out on his maternal grandfather's cattle ranch when he wasn't in school. He attended college for most of a year before being drafted by the army at age 18. He served in the infantry in Europe, part of that time under General George S. Patton. Returning to Oklahoma after the war, Magness attended and graduated from Oklahoma State College in Weatherford. He married Betsy Preston, from a large cotton-farming family. The same week, he went to work at Anderson Clayton, the world's largest cotton company.

A small-framed man with an easy grin and a shy manner, Magness hung around cotton gins, striking deals to buy seeds for processing into salad oils, margarine and the like. "Everybody paid about the same price so [the ginner] usually sold to people he liked the best and who he thought would treat him the best," Magness recalled. "That was a very pleasant time in my life because I'd carry a set of golf clubs, a fishing pole, a jug of booze and whatever the ginner wanted to do, we did."

A chance meeting one night at a cotton gin tipped Magness to cable. Two men walked in who had blown a rod in their pickup truck. Magness gave them a ride back to Paducah while they told him about a cable system they had built there. "So I listened and thought a little bit about it, and about a week later, I had thought enough about it, so I went down and looked them up and talked to them some more," said Magness. "About 30 days later, we were stringing wire."

Bob and Betsy Magness sold their cattle, borrowed some money from a bank, plus $2,500 from Magness's dad, and built their first cable system in Memphis, Texas. The franchise was approved with the help of the town's mayor, who happened to be Magness's banker and hunting partner. Betsy Magness did the company's books at their kitchen table, with the couple's two sons, Kim and Gary, underfoot.

Magness bought equipment, as did most of the early cable guys, from Jerrold Electronics, a Philadelphia company headed by Milton Jerrold Shapp, a future governor of Pennsylvania. Jerrold used its leverage to take stakes in fledgling cable companies, and charged some operators 25 cents per month for every subscriber signed up. "If a system is operated properly, it's like shooting fish in a barrel," said Schapp in 1953. Schapp ended up in the barrel himself four years later when the U.S. Department of Justice filed an antitrust suit against Jerrold, forcing divestiture of its cable ownership.

At that time, cable revenues were plowed back into building systems and doing deals. There was little resistance from broadcasters and government agencies. Cable TV was a novelty and a public service, delivering ad-supported TV programs to rural viewers who wouldn't see them otherwise. The cable guys were stealing free TV signals and sending them down a wire for a price.

For the most part, that served everyone's purpose.

Until it didn't.

—

TO THRIVE, THE CABLE gang needed to deliver more and more channels, and they went farther and farther out of market to get them. Microwave relay stations were used to bounce signals across the country. Daniels and Schneider used a microwave relay from Denver to Casper. Magness carried five channels from Salt Lake City to Bozeman, Montana. With a tower on Carter Mountain, Roy Bliss tried to deliver a second TV signal from Idaho Falls, Idaho, to subscribers in Worland.

The broadcasters slowly woke to the fact that they were supplying programming for a new TV delivery system over which they had little control. Dan Shields, an official of the National Association of Broadcasters, gave a pretty good description of this cutthroat situation to a meeting of Georgia broadcasters in 1962. "What we have here is a completely unregulated business competing against a regulated industry, using as its major weapon the very product which its competitor turns out, and paying nothing for the product," he said.

Long a force in Washington, whose denizens needed access to radio and TV airwaves to campaign back home, the powerful broadcasting lobby tried to strangle cable in its crib.

The FCC was enlisted to block cable's use of microwave to import distant broadcast signals, starting with the Carter Mountain case in 1962. Lawyer-to-lawyer combat reached the U.S. Supreme Court, with the debate generally resolved in the 1970s when new rules allowed broadcasters to charge copyright fees.

"The broadcasters wanted it to be the three networks and nothing else," said Gene Schneider, whose company, United International Holdings, ended up operating entirely overseas. "They fought anything that veered from that. Those guys own the FCC and still do. How did we prevail? Because we were right, because we had a product that people wanted. But they fought it every step of the way."

Daniels put it another way. "I have often said that the two friends we have had since day one have been Main Street and Wall Street."

One price paid by viewers to this day from the regulatory jousting is that cable companies must carry the signals of every local broadcaster, regardless of programming or viewership. Congress and the Supreme Court have upheld this "must-carry" rule over the objections of the cable gang that it violates their free speech rights.

As cable captured nearly one million subscribers in the mid-1960s, broadcasters grew paralyzed by fear of pay-TV services that would divert viewers and revenues. Of particular concern were renegades like Irving Berlin Kahn, a portly son of Russian immigrants named after his famous songwriting uncle, Irving Berlin.

Kahn became an unlikely drum major for the University of Alabama's Million Dollar Band and even penned a football fight song, "Rootin' for the Rose Bowl." He left a cushy executive job at Twentieth Century Fox movie studios to found TelePrompTer Corp., a New York company that sold rolling text displays used by public speakers. TelePrompTer moved into cable and pay-per-view events, delivering the 1960 title fight between Floyd Patterson and Ingemar Johansson to 25,000 cable subscribers, who paid $2 apiece.

In the mid-1960s, Kahn was convicted of bribery and perjury charges after he funneled $15,000 to associates of the mayor of Johnstown, Pennsylvania, to try to keep TelePrompTer's cable franchise exclusive. He served less than half of a five-year prison term at "Club Fed" facilities at Allenwood and at Eglin Air Force Base in Florida, where he devised the facility's cable system. Soon after his release from federal custody, the buoyant Kahn addressed a Texas cable convention with this opening line: "Now, where was I before I was so rudely interrupted?"

Broadcasters claimed that charging viewers extra for specialized programming, as Kahn did for boxing matches, would cannibalize ad-supported free TV. Facing competition, they exhorted each other "to strangle it, stifle it—or seize control of it," according to a 1964 alarm bell sounded in *Broadcasting* magazine. Their argument to protect free TV was both heartfelt and self-serving. "America does not need a class system in television based on the ability to pay," argued the National Association of Broadcasters president Leroy Collins, adamant behind black horned-rim glasses.

Had the broadcasters been less myopic, they might have become partners with cable and made themselves invaluable as dominant program suppliers in the coming decades. Instead, they fought losing battles against cable, or tried to co-opt it, and ended up a declining power in home entertainment by the late 1990s.

Broadcasters had been granted use of the public airwaves to make money, but they faced heavy regulation in return. They wanted similar restraints on their upstart cable competitors.

In California, one million petition signatures were collected by a group called the Citizen's Committee for Free TV in support of a 1964 ballot question to block any pay-TV service beyond basic cable. The ballot measure, which passed, was a thinly veiled attempt by broadcasters and theater owners to undermine Subscription Television Inc., a company that intended to sell pay-per-view Los Angeles Dodgers and San Francisco Giants baseball games, as well as movies and performances. Pat Weaver, the president of Subscription Television, eventually won a state supreme court ruling overturning the law, but his company died during the delay. Early on, Weaver challenged Collins and the broadcasting lobby to admit "that in America the marketplace is the arena in which decisions affecting innovations would be made, not in political pressure arenas, and not by monopolistic intrigues or the unholy alliance of some motion picture theater owners and some broadcast owners so obviously affected in their opinions by our new competition."

Within three decades, a similar "unholy alliance" would form between the cable industry and its programmers when satellite TV and other competitors tried to get in the game. The cutthroat game never changes. Only the players do.

While broadcasters feared cable and pay-TV services in the 1960s, some of them tried to get in on the action, buying cable systems from Wilmington, North Carolina, to Santa Barbara, California. President Lyndon Johnson caught the fever. His family owned 84 percent of Texas Broadcasting, which controlled the only television station in Austin. Texas Broadcasting had an option to buy 50 percent of Capital Cable in Austin, a deal approved by Johnson's FCC. An impertinent questioner on a syndicated TV show, *Youth Wants to Know*, asked FCC chairman William Henry whether the Johnson family deal was "an attempt to monopolize television rights" in Austin. With bureaucratic aplomb, Henry replied that the proposed deal dated back to 1957 and "speaks for itself." Cross-ownership rules later restricted broadcasters from owning cable companies going forward.

Broadcasters even tried to launch their own pay-TV efforts, which were wildly unsuccessful. Teleglobe tested a service, backed by Bill Daniels. A first-run movie was shown between 9 P.M. and 11 P.M. on a local broadcast channel. But viewers had to pay $3.50 to get the sound piped in over a phone line. That experiment, like so many cable forays in later years that boasted video-on-demand, interactive TV services, 500 channels and the like, was an ambitious attempt that failed to fulfill the hype. Many sunk without reaching a mass market or showing a profit.

Basic cable, however, was undeniably a hit, capturing nearly four million subscribers by 1970. What kept cable thriving through the next decade was a pay channel that delivered—for the first time—uncut, uninterrupted and commercial-free movies direct to living rooms. Home Box Office forever divided cable from broadcast TV in viewers' minds. It also fed demand for cable in the cities, which typically had good broadcast reception and therefore little need for a duplicative television service.

Originally named the Green Channel by founder Charles Dolan, HBO battled the Hollywood studios, broadcasters and the FCC to obtain movie rights. It later lost viewers to videotape rentals. But early on, HBO's biggest hurdle was churn from subscribers who signed up for a month or two, then dropped the service, unwilling to pay extra for programming like the 1973 Pennsylvania Polka Festival.

HBO participated in another little-noted event that year. The channel, along with TelePrompTer and Scientific-Atlanta, demonstrated a live, point-to-point satellite program, linking House Speaker Carl Albert to a cable industry convention in Anaheim, California.

For years, broadcasters and phone companies had used satellites—also called birds—to beam TV signals and phone calls overseas. In the mid-1960s, the big three broadcast networks used Comsat's *Early Bird* to beam live transatlantic TV signals, including a CBS *Town Meeting of the World* on the Vietnam War that linked panelists in New York and London. AT&T and its counterparts around the world, which typically relied on undersea cables to carry international calls, paid $4,200 a month for a satellite feed on *Early Bird*.

Domestic cable signals, however, were bounced from microwave tower to microwave tower in an expensive and sometimes spotty relay. Or, videotapes were shipped across the country in a procedure known as bicycling. Using satellites to deliver cable programming within the United States was like discovering a new use for the sun.

Gerald Levin, a former divinity student and attorney who took over HBO in 1972, gets much of the credit. He convinced HBO's parent company, Time Inc., to spend $7.5 million to lease one channel for six years on an RCA satellite due to be launched in December of 1974. Negotiating with Don King, the boxing promoter with the bride-of-Frankenstein hairdo, Levin licensed the "Thrilla from Manila" heavyweight title fight between Muhammad Ali and Joe Frazier, which took place on September 30, 1975.

Levin's idea was to put the fight on satellite and beam it as an HBO production to cable systems. This required a big investment, not only a satellite in orbit, but dish receivers on earth that were 10 meters, or 33 feet, in diameter. The dishes cost $100,000 apiece and only a handful of cable operators had them. On fight night, Levin and HBO hosted a party at a Holiday Inn near Vero Beach, Florida. Onto the hotel's TV screen came live the championship bout from half a world away. Ali won in 15 rounds. Levin won acclaim by linking the physics of satellite to a cable industry poised to go national.

On the programming front, HBO grew to become cable's premium pay channel, paving the way for regional sports-and-movie services like Dolan's Madison Square Garden network, as well as national pay channels like Showtime, the Movie Channel and Encore. On a technical level, HBO's satellite delivery was eventually aped by every major and minor cable network. Pat Robertson's Christian Broadcasting Network went up on a bird. So did the Entertainment and Sports Programming Network (ESPN), Music Television (MTV) and the Cable Satellite Public Affairs Network (C-SPAN).

From sleepy Atlanta, Ted Turner turned Channel 17—call letters WTCG for Watch This Channel Grow—into WTBS, the Turner Superstation, a southern stew. By day, *Cartoon Cavalcade* and *The Munsters*. By night, Burt Reynolds movies, Atlanta Braves and Atlanta Hawks games. Turner connected with a man named Ed Taylor, whose company, Southern Satellite Systems, in Tulsa, Oklahoma, began delivering WTBS nationwide by satellite and eventually collected 10 cents a subscriber per month from cable systems. Meanwhile, Turner began luring national advertising accounts like Toyota.

It was a new world. But the cable gang's euphoria over satellite technology to deliver national programming services hid a disquieting reality. If cable company dishes could receive satellite signals, then so could a dish in someone's backyard. And if that was possible, why have a cable middleman at all?

—

ONE DAY IN 1976, a Stanford University professor named Taylor Howard heard a curious thing from a graduate student analyzing weather data from an Intelsat satellite. A signal picked up from another bird had the characteristics of video. Howard, 44, pondered how he might capture the signal. At the time, he was an electrical engineering professor, amateur radio buff and lead scientist on several NASA interplanetary probes.

There was no cable TV service at his ranch in the Sierra foothills east of Sacramento. He had no idea that HBO even existed. Howard, his wife and their five kids watched broadcast TV captured by a 60-foot tower on their property. In the backyard was a 15-foot aluminum dish, once owned by a telephone company for the purpose of receiving long-distance calls by microwave. Howard bought the dish for scrap, planning to refurbish it and use it to bounce Morse code signals off the moon.

Moonbounce is a little-known, low-cost and low-speed method of satellite communication that uses the moon as a natural satellite orbiting Earth. A signal bounced to the moon returns to earth in about 2.5 seconds—not very useful for real-time voice or data transmission. The moon's craggy and pitted surface is also less than ideal for reflecting clean signals. The U.S. Navy experimented with moonbounce in the 1950s, but abandoned it.

Howard decided that his 15-foot scrap dish might be used to capture the mysterious video signal. Using spare electronic parts, Howard built a low-noise amplifier for the focal point of the dish. He connected it with coaxial wire to an FM receiver and then to a 19-inch Sony Trinitron. The entire setup was built over four months at a cost of less than $1,000. The tense moments were more domestic than technical. "When I wanted to experiment, the TV had to be uninstalled from the living room and carried out to the garage," said Howard. "There were lots of negotiations about timing on that."

On September 14, 1976, Howard manually pointed the dish to the azimuth, or angle, from which he believed the video signal came from the sky. "It was harder than hell to find," he recalled in a 1998 interview. "There was only one signal, the width of the antenna beam was less than a degree and I wasn't exactly sure of the frequency. It was kind of a four-dimensional search because I also didn't know what time the signal came on the air."

After some adjusting, a scratchy red picture with fuzzy white lettering appeared on the screen, like a mutant message from outer space. It read, "Attention all earth stations," and listed satellite coordinates. The coordinates were intended to track a satellite that delivered the HBO signal, which cable companies would then capture, send down their lines and sell to subscribers.

With his home-built dish, Taylor Howard joined the party for free. The TV industry, about to be rocked by HBO, unknowingly faced another seismic event from the likes of Howard.

"I kept very quiet about it," he said. "I was concerned about my amateur and commercial radio licenses. I didn't want to lose them." Howard wrote a letter to HBO in New York, telling them that he enjoyed watching their movies from a home satellite dish. "I understand it's a pay service," he wrote in pencil on a piece of yellow legal paper. "I would like a monthly subscription."

He never heard back.

Howard's invention attracted the attention of Bob Cooper, who visited Howard's home in May of 1979. Seeing that Howard kept lab notebooks on his work, Cooper suggested they publish his findings for other hobbyists who wanted to build home satellite receivers. At $40 apiece, the 31-page *Howard Terminal Manual* sold out its initial printrun of a few hundred copies, and later sold thousands more copies. Cooper conducted a seminar and sold the manual at South Oklahoma City Junior College.

The buzz was building. In 1979, the FCC lifted requirements that satellite dishes had to be licensed and 33 feet in diameter or larger. On the cover of its 1979 Christmas catalog, Neiman Marcus featured "the first remote-controlled, multi-satellite antenna" dish from Scientific-Atlanta. It cost $36,500 at retail, but the catalog copy made it sound like a steal.

> With a flick of the switch you have options that include: over
> 2,000 sporting events yearly, 6,000 hours of specialized children's
> programming, 10,000 top movies, live nightclub shows from Las

Vegas and New York, sessions of the U.S. House of Representatives, a direct line to news agencies and business reports, the new "super stations" and much more.

A 1979 *Forbes* article on the pay-TV industry pictured none other than John Malone, his left hand resting atop the lower arc of a large, snow-covered satellite dish in his backyard. Dressed in a turtleneck and a boxy sports coat with kangaroo pockets, Malone had an exultant look on his face, with his right index finger pointed to the sky. The article had just one quote from Malone, but it was succinct: "I can watch a dirty movie any hour of the day or night."

The big backyard dishes went by a variety of names: earth stations, C-Band, TVRO, for Television Receive Only, and BUD, for Big Ugly Dish. The big dishes captured all sorts of video signals from an array of satellites. From the orbital ether came the 1980 Moscow Olympics from the Russian *Molniya* satellite. A key attraction were backhauls—live feeds bounced from mobile news trucks to satellites, then back down to TV stations for editing and broadcast. Lounging in a La-Z-Boy recliner and rotating the backyard dish by remote control, a viewer could intercept the feeds and observe TV reporters and interview subjects biding time, primping or talking to people behind the camera. One backhaul showed former CIA director Stansfield Turner making phone calls in a *Nightline* studio. Another showed Max Robinson, a co-anchor of the ABC evening news in the 1980s, being approached by a man with a can of hair spray during a studio break.

"What is this shit?" asked Robinson.

"It's hair spray," said the man.

"Hair spray," said Robinson. "You honkies don't know nothing about hair spray."

Backhauls were typically odd and unscripted, yet often compelling, because they subverted the staged, scripted and frosted newscasts seen by most of the viewing public. It was the video equivalent of wiretapping. Comedians including Johnny Carson,

Jay Leno, Roseanne Barr and Mort Sahl were among those who eventually bought big dishes and monitored backhauls for their own amusement and potential material.

Big dishes were the precursor to mini-dish satellite television provided by the likes of DirecTV and EchoStar. But it would take the decade of the 1980s for technology and economics to make that a viable mass-market service.

During that time, Malone and the cable gang achieved a lucrative chokehold on pay television by remaining the primary gatekeeper and toll collector. Cable's growth in the 1980s tracked perfectly with a voyeuristic, consumer culture hungry for channel choices and weary of the stodgy broadcast networks. Home owners took down aerial antennas from their roofs in favor of black cable lines that snaked through their living rooms, bringing everything from the Discovery Channel to *Midnight Blue*, a New York–based show that billed itself as the "60 Minutes of sex."

Cable started out delivering broadcast signals to rural towns. It flourished by creating its own national networks. And it became a fearsome monopoly by locking up agreements to serve the nation's cities. Cable franchising was the pro wrestling of the business world—a showcase for muscle, flash and rigged outcomes.

"Virtually everybody but TCI went tearing into the big cities," said Stratford Smith, the onetime cable industry attorney. "There were instances where franchising authorities assured the cable operators that they would get one of the franchises if they included certain people as stockholders. It was really a form of bribery. There was quite a bit of that. It was almost standard practice in the major communities."

While patronage and political fixes are old hat, the boldness of the cable carpetbaggers and the kickbacks solicited by local fixers were exceptional. Most franchises were exclusive. Or, if the city territory was divided, the winning cable companies controlled separate turf. A franchise typically lasted 15 years and promised to be a cash gusher, with 5 percent of the flow diverted into city coffers. City officials negotiated for public, educational and government programming, called PEG channels, which were typically under-funded and little watched. Cable company hopefuls made often ex-travagant claims about advanced, interactive and educational serv-ices to help win the bids, while funding everything from private charities to swimming pools.

"In 1983, along with Cablevision of Long Island, we bid an in-teractive system for the city of Sacramento that featured e-mail as a basic service," wrote Bill Frezza, a onetime manager at General In-strument Corp. "Home banking, home shopping, interactive infor-mation retrieval, multiuser games—we demonstrated them all. Luckily for Sacramento, we lost to a competitor that agreed to plant 20,000 trees. The trees were real. Two-way cable turned out to be a fantasy."

To win Boston, Charles Dolan's Cablevision offered every city resident the chance to buy up to 25 $1,000 cable bonds, with a guar-anteed rate of return. Critics called the resulting franchise agree-ment "an orgy of excesses."

Warner Amex, a partnership of Warner Communications and American Express, won a 15-year franchise in Pittsburgh in 1980 by ingratiating themselves with the local power base. They doled out 20 percent of the business to the Urban League of Pittsburgh, the United Negro College Fund and 15 other predominantly black local groups. "When politics are involved, it's never too late" to get a piece of the action, said a black businessman in Pittsburgh at the time. "And politics are very much involved in cable TV."

The Warner Amex bid beat out three contenders, including American Television & Communications Corp., a subsidiary of Time Inc. ATC even offered to provide a basic cable package to the city's 170,000 homes—for free. When Warner Amex won, ATC filed suit against its rival and the city, claiming the bid process was rigged. That coincided with a federal probe into bribery and influ-ence peddling. Neither the civil nor the criminal case proved par-ticularly fruitful.

Yet, the template for winning a franchise was struck. Bidders doled out ownership stakes to local power brokers and promised the moon. The winner would hold on tight while the also-rans and good-government types cried foul and sometimes called in the feds. The five winning bidders who divvied up Houston's cable franchise in 1979 were companies represented by, respectively, the mayor's former campaign manager, a former city attorney, another former city attorney, a political fundraiser and the mayor's personal attorney. The two losing bidders were represented by the chairman of the state Democratic party and yet another former city attorney.

It was a grubby, speculative land grab. Within eight months, three of Houston's winning bidders had virtually sold out their companies at a profit to the two that remained, Warner Amex and Storer. One of the unsuccessful bidders sued. After a five-week trial, a federal jury found that the city, its mayor and one of the local cable fronts had "participated in a conspiracy to limit competition."

Robert Sadowski, a consultant to Houston during the franchise process, said he lost his job for not rubber-stamping the greasy deals. "If ever there was an argument for an increase in federal oversight in the cable area, Houston would be a strong reason for it," he said later.

The feeding frenzy continued.

The Cincinnati School Board offered to accept a 20 percent stake in any prospective local cable company, apparently seeing no conflict of interest in both sanctioning, and angling to profit from, a service intent on luring young eyes from books to the small screen.

In Denver, 22 individuals connected to the mayor and city council were brought in as local partners to a 15-year franchise won by Mile Hi Cablevision, led by broker Bill Daniels. The lucky 22 were dubbed "instant millionaires" in the press. "The best offer doesn't necessarily always win, and we may not be willing to do things that may be required to win," said Daniel Ritchie, then the chairman of Westinghouse Broadcasting, which lost its bid to wire Denver for cable.

Trygve Myhren, the president of ATC, said he was so fearful of bribery scandals that he sometimes hired private detectives to investigate his own company's local contacts. The concern grew grave in St. Louis. "I really started to worry about the safety of our people," said Myhren. "One of the reasons we pulled out of the St. Louis bidding was that two people related to our local ownership groups died under suspicious circumstances."

The franchising process was "helter-skelter," according to Thomas Wheeler, the cable industry's lobbying chief. He told Congress in 1983 that "a consumer will have a couple choices of cable companies. There will be two cable wires running down the street."

Not a chance. The cost of building cable systems from scratch and the local influence peddling conspired against it. The cable franchising process was ultracompetitive in the bidding and monopolistic in actual service. "The municipal franchise is issued not to regulate a monopoly but to create one," wrote Thomas Hazlett, who later became chief economist at the Federal Communications Commission. "It is simply a legal barrier that allows cities to treat cable as an urban pork barrel."

There was a price to pay, however, for all the quick fixes and wild promises. Warner Amex won Dallas with a blue-sky bid in 1980: 100 channels, with 18 devoted to public access, plus 10 community studios and a private data link for businesses and government offices. Warner Amex wired downtown Dallas first, but found that few businesses subscribed. Meanwhile, apartment-house owners contracted for rooftop satellite dishes that could serve all of a building's tenants. This service, with the unappetizing acronym of SMATV—for satellite master antenna television—did not require city franchise approvals because the wiring did not cross public rights-of-way. Racking up $60 million in losses in five years, Warner Amex sold the Dallas system to Heritage Communications. "We felt a new name would do much better," said Drew Lewis, the Warner Amex chairman.

Unlike Time Inc. and Warner Communications, which later merged, TCI didn't have the deep pockets, desire or political skills to play the urban franchise game.

"We weren't going to build libraries, give away free fire engines and send the mayor's 18 illegitimate kids to college," said Malone bluntly. He was, however, eager to pick up the pieces when a deal went bad. TCI bought the Pittsburgh franchise in 1984 for $93 million, $7 million less than Warner Amex poured into it.

Magness said later that corruption in franchise bidding "was probably one of the blacker parts" of cable history and that TCI walked away from several deals "when the midnight phone call" came for a payoff. "It wasn't extremely prevalent, but it went on enough and the bad thing about it is that you never really knew because it was a bag man somewhere, an attorney, or something of that nature. You don't know that he wasn't out there just to feather his own nest. He may have never intended to give anybody a nickel."

One call for draining the franchise cesspool came from Warren Buffett, the Nebraska newspaper publisher, better known as the Berkshire Hathaway billionaire investor, free-market maven and buddy of Bill Gates. In a 1980 *Washington Post* column, Buffett noted that the Sun Newspapers of Omaha, which Buffett ran, were offered a 20 percent stake in a local cable franchise if the newspapers helped win a contract from the city. "What would be blatantly illegal if only two parties were involved—the grantor and the grantee—apparently can be legalized if not deodorized by the presence of three parties—grantor, free-riding party of influence and grantee," Buffett wrote.

There were myriad calls for reform and regulation of the cable industry. But as the urban franchise wave swelled in the late 1970s, then subsided by the mid-1980s, cable caught a big break during the deregulatory reign of President Ronald Reagan. The 1984 Cable Act lifted local cable price controls as of 1987, which freed investment in new programmers and helped fuel cable's subscriber boom throughout the decade. The legislation was shepherded through Congress by Representative Tim Wirth, a Democrat of Colorado, who was sometimes known as the congressman from TCI for his pro-cable stance.

Coming the same year as the court ordered breakup of AT&T's local phone monopoly into the Baby Bells, the 1984 Cable Act also prevented the local phone companies from owning cable TV companies in their own regions as an anti-monopoly provision.

Yet, the act simply entrenched the cable gang's budding monopoly. Had they gone slow, taken modest price increases and built good relationships with their subscribers, cable might have grown steadily and favorably into the future. Yet, some in the cable gang abused their newfound freedom by jacking up rates, skimping on service, playing hardball on Main Street and dismissing the thunderclouds building back in Washington.

The Colorado cable cowboys led by Malone wore the black hats.

FOUR

THE CHESS
MACHINE

J OHN MALONE'S FAMILY had the first television set in Milford, Connecticut. It was a six-inch, black-and-white model that drew a crowd to their house on tree-shaded Gunn Street in the mid-1940s. The boy was as interested in how the tube worked as in what it displayed. He later tinkered with TV sets in a backyard barn and rehabilitated surplus General Electric radios, reselling them for $7 apiece. He helped his father, Daniel, fix up foreign sports cars.

Daniel Lafine Malone was a manager in the electronics division of General Electric for 30 years. He worked at plants in Bridgeport, Connecticut, and in Schenectady and Syracuse, New York, which required travel that kept him away from home during the week. He was quiet, later described by his son as "an intellectual with white socks."

His wife, Jule Custer Malone, was a Philadelphia native who attended Temple University and taught elementary school in Stratford, near Milford. She was as outspoken as her husband was not. "Whatever was on her mind, she spoke her piece," said Edward Kozlowski, a former mayor of Milford who lived on Gunn Street for more than 70 years and knew the Malone family growing up. "She didn't pull her punches. And it was too bad if you didn't like it."

It was a quality her son would inherit and amplify.

Blessed with brains and athleticism, Malone separated himself from the herd early on. He was repelled by graveyard headstones in his hometown. "Nothing distinctive. Nothing unique," he said. "So I've always wanted to be different, always wanted to be unique." He attended public schools in Milford, then transferred to Hopkins, a private prep school in New Haven that took only day students. Named after Edward Hopkins, a former governor of the Connecticut colony, it was founded in 1660 as a one-room schoolhouse committed to "the breeding up of hopeful youths."

Malone was a math whiz who blazed through calculus classes. He also fenced, played chess and soccer and held the school record in the discus. His intensity was evident early on. Next to each photo in Malone's senior class yearbook was the student's nickname and a hand-drawn illustration. Malone's nickname was The Bat, accompanied by a black-ink drawing of a winged creature hanging upside down from a branch.

Along with 13 of his classmates, Malone attended nearby Yale University, founded 41 years after Hopkins. At Hopkins, they say Yale was built so Hopkins students wouldn't have to travel to Harvard for an Ivy League education. Malone received a merit scholarship and majored in electrical engineering at Yale. He later disdained much of the faculty and his classmates as "socialistic."

Fred Andreae, a freshman roommate of Malone's who became an architect, said that Malone ran track and occasionally performed one-arm pull-ups on a bar across a doorway in their dorm room. He recalled Malone "as not having much money and hanging around with a rough crowd in Milford."

For the most part, Malone kept to himself and his studies, spending much of his free time in Milford, where he courted Leslie Ann Evans, a secretary at the police department. They had been introduced on the beach as teenagers by his older sister, Judy. Shortly after Malone graduated from Yale in 1963, he and Evans married. They later had a daughter, Tracy, and a son, Evan.

"My background is German and Scotch-Irish," Malone once said. "My wife says that makes me bullheaded, cheap and alcoholic. I prefer to say it makes me efficient, productive, economical and, perhaps, glib."

He went to work as a systems engineer at AT&T Bell Telephone Labs in Holmdel, New Jersey. AT&T helped finance his three advanced degrees. Malone earned a master of science in industrial management from Johns Hopkins University in Baltimore in 1964, a master of science in electrical engineering from New York University in 1965, and a doctorate in operations research from Johns Hopkins in 1967.

His thesis papers illuminate Malone's mind-set in an era when some of his generation waged war in Vietnam, protested in the streets at home or grabbed a rung in middle America. Malone focused on "Price, Production and Inventory in a Monopolistic Firm," the title of his master's thesis at Johns Hopkins. This work fit nicely with the concerns of Ma Bell. Similar issues would play out in the rural cable industry that Malone would enter and transform.

His doctoral thesis, "Traveling Salesman Algorithms: A Theory and Its Applications," revealed Malone's conceptual powers. The title refers to a deceptively simple question famous among mathematicians: What is the shortest total traveling distance among many points? The complex of variables, including number of destinations, distance between them and cost to get there, makes the number of outcomes potentially infinite. Malone crunched it.

"John thought through a dimension that most people had never thought of," said Professor Mandell Bellmore, a faculty adviser to Malone at Johns Hopkins. "He understood why a computer will die trying to solve certain problems. He made a major contribution." Another faculty adviser, Professor George Nemhauser, recalled Malone's laser focus.

"Given that it was a time of social upheaval and Vietnam, the students tended to be more idealistic," he said. "John wasn't one of them. He had this business attitude about getting his Ph.D. There was a custom at Hopkins that when someone defended their Ph.D. thesis, a group always got together to spend an evening drinking beer down at The Block in Baltimore, this sleazy area where all the strip joints were. I don't remember John ever coming to one of those things." Malone could have become a prominent professor of higher mathematics, both Nemhauser and Bellmore said. Instead, he wanted to make his mark in business.

"What do people care about more than anything in the world?" Nemhauser recalled Malone asking one day. "Well, maybe one is TV, this new thing starting out called cable TV. After a couple of years at Bell Labs, finishing up my obligation, that's what I'm going to get into."

In 1968, after completing his Ph.D., Malone began working at McKinsey & Company Inc. in New York as an associate management consultant. He traveled widely, advising companies like IBM and General Electric. He joined one of McKinsey's clients, General Instrument Corp. (GI), in 1970 and became president of Jerrold, its cable TV equipment division. GI chairman Moses Shapiro became a mentor, offering what Malone later called the best business advice he ever received: "Son, this is Talmudic wisdom. Always ask the question, 'If not?' Few people have good strategies for when their assumptions are wrong."

Malone would soon get a chance to test "If not." Shapiro passed up Malone in favor of a man named Frank Hickey as GI's chief executive officer. By that time, Malone knew the major players in the cable industry. Steve Ross, a onetime funeral director and corporate showman who ran Warner Communications in New York, offered Malone the company's top cable job and a chance to remain on the East Coast.

Instead, Malone connected with Bob Magness in Denver, who had formed TCI in 1968 from a group of small cable systems. When Malone flew out on sales calls as Jerrold's president, he stayed at the home of Bob and Betsy Magness. "They treated me as part of the family, not as a salesman," Malone recalled.

The Magnesses' own two sons, Kim and Gary, were teenagers in the early 1970s. Kim later became a TCI board member, but neither son took up the cable business, preferring to help manage the family's cattle and ranching properties. As late as 1989, with Kim and

Gary in their mid-30s, Bob Magness said, "We are really attempting to find the thing that these boys want to do for the rest of their lives."

That wasn't an issue for Malone, who found his mission with Magness. TCI had a deal with Jerrold to buy equipment at the lowest possible rate—a "most favored nation" clause that became TCI's principal leverage with vendors. Yet, one particular cable system that Magness bought had a better deal with Jerrold. Malone had the uncomfortable task of explaining why and making amends. It was that kind of bargaining that forged their partnership. Magness asked Malone to join TCI in 1972, which he did, moving his family to Denver. Thirty-two-year-old Malone became TCI's president and chief executive officer in April of 1973 at a salary of $60,000, half of what he made at Jerrold, but with a ground-floor opportunity.

The power and perils of Malone's leadership were revealed early on. He helped secure $100 million in loans in 1973 from a group of banks, including the Bank of New York, to help TCI compete in the microwave transmission business with AT&T and Western Union. In November of that year came a showdown with Vail, Colorado, 100 miles west of TCI's headquarters in Denver.

"Vail was born in a sheep pasture at the foot of an unnamed mountain, where a two-lane road meandered along Gore Greek past a few sagging barns and homestead cabins," wrote author June Simonton. U.S. Army ski troopers of the 10th Mountain Division trained in the area prior to World War II. One of them, Peter Seibert, returned and joined with local businessmen to create the ski resort named Vail, which opened for business in 1962. It quickly earned a world-class reputation and counted its one millionth skier visit in 1968 on its way to becoming the Alps of the interstate and a winter playground for the wealthy.

The village of Vail was another story, economically reliant on the resort, but with an eclectic mountain culture. "When religious groups began fund-raising for a chapel, a bartender at Donovan's Copper Bar put jars of his own homemade pickled eggs out on the bar, labeled them 'Pickled Eggs for God' and raised enough money to donate a pew," Simonton wrote. Failing to meet the state residency requirements of 100 people living within a square mile, the town did not incorporate until 1966. The mayor, John Dobson, wrote and produced one-act melodramas such as *Dirty Doings at the Depot* under the pen name Gregory Beresford Skeffington.

When TV came to the valley, the mayor and council moved their weekly Monday night meeting to Tuesday night so they could watch Monday Night Football on ABC. At an elevation of 8,000 feet and blocked by mountains, Vail had minimal broadcast TV and radio reception. Four TV channels were relayed from Denver by microwave and then through cable lines to subscribers. The picture quality was often poor and maintenance shoddy. One longtime resident recalled cable lines strung along the ground, exposed, like an extension cord. TCI bought the system and proposed a long-term franchise renewal in 1973 at the same time that the Vail council voted to end it.

For cable companies, a franchise was life or death. Years after Irving Berlin Kahn served prison time for bribing officials to keep his Johnstown cable franchise, he offered this explanation: "If we lost this goddamn thing, it would be a major loss. How would you feel about losing a kid of yours?"

TCI was just as territorial but with a more surgical style—courtesy of Dr. Malone.

One Thursday night in mid-November of 1973, cable TV screens in Vail households suddenly went black. Up popped this message: "If you have any questions about cable TV service, contact John Dobson and Terrell Minger," naming the mayor and town manager. The home phone numbers of both men were listed on the screen. Minger hurried over to Dobson's house to help answer the phone, which rang off the hook, as it did at Minger's house and at the police station. The cable blackout continued through the weekend. Minger and Dobson answered hundreds of calls from furious residents, some of them particularly wrenched about missing the Sunday football game between the Denver Broncos and St. Louis Cardinals televised from Busch Stadium. The city and TCI went into

24-hour negotiating mode. By Monday, service was restored. On Tuesday, the council and TCI came to terms at a meeting attended by 100 residents wanting to know what the hell was going on.

"I realized this was a pretty heavy-handed company," Minger recalled 25 years later. "They tried to hold the franchise hostage and they did it when it would have maximum effect, with people wanting to watch the football game. I think in retrospect it kind of hurt them. Bullying a small town is probably not the way to win friends and influence people."

Malone saw it in purely rational terms. "We were demonstrating that the citizens wanted the cable service," he said years later.

An internal TCI memo circulated among the company's executives at the time stated, "The ensuing publicity in the trade not only added stature to our image but could possibly serve as an example to other operators with subsequent benefits to our entire industry."

Malone later claimed to have disagreed with that assessment. "I didn't want the reputation of having to use bare knuckles in order to get the cities to listen to reason," he said. Yet, that's exactly the reputation TCI did cultivate. Other members of the cable gang may not have employed the same tactics, but neither did they publicly disavow them.

"There was a reluctance for any of them to jump on TCI publicly, in some part, because none of them were absolutely clean," said Stratford Smith, the cable industry attorney. "They were all fighting for their lives. If you jump too hard on a TCI, all of a sudden you don't have TCI's dues in your trade organization."

TCI backed down to no one. While Malone bought cable systems and kept bankers at bay, a cable veteran named J. C. Sparkman ran TCI's field operations on a shoestring, with a particular antagonism to labor unions. Cable companies, unlike phone companies, were not heavily unionized. Sparkman aimed to keep it that way at TCI. One night in the mid-1970s, as TCI faced labor strife after buying a cable system in California, Sparkman walked toward his car in a Denver-area restaurant parking lot. He was jumped by two men and beaten to a bloody pulp.

"My face looked like hamburger," Sparkman recalled years later. "I had to have my forehead rebuilt." No arrests were made, though Sparkman suspected it was union thugs. "I couldn't allow that to stop me," he said. "I took on unions with a vengeance. I decertified as many as I could. I would not give in to their demands."

Capital-intensive, heavily indebted cable companies struggled to survive in the battered U.S. economy of the 1970s. They borrowed money, Malone said later, "at rates the Mafia would be ashamed to charge."

One day in 1976, lenders from the Bank of New York and Philadelphia National Bank flew out to TCI's one-story headquarters and lowered the boom. TCI was in technical default on a loan agreement. The bankers wanted to raise the interest rates to reflect the higher risk. Malone took a set of keys out of his pocket and tossed them on the table. "I'm doing everything I can to run this company," he said. "You can put us in the tank. You can take the company over. You can run the company if you want. But you can't raise the interest rates unless we agree. And we're not agreeing."

The bankers left without announcing a decision. Magness's blood pressure was so high that he couldn't stand to be in the office the next day when the Bank of New York official called. "You guys misinterpreted what we were saying," the banker told Malone. "We were saying we should raise the interest rate but we're not going to in light of the hard work you guys have done." TCI dodged the executioner's blade and Malone later joined the Bank of New York's board of directors.

A key financier of fledgling communications companies in the mid-1970s through the 1980s was Drexel Burnham Lambert, the junk bond firm masterminded by Michael Milken in Los Angeles. He encouraged businesspeople to tell him their dreams. Some of the dreamers were corporate raiders who used junk bonds to buy enough stock to put a company in play, take it over or demand payments to walk away. Some of the dreamers were entrepreneurs like Rupert Murdoch, Ted Turner, John Malone and MCI founder Bill McGowan, all of whom had difficulty getting traditional bank loans. "We did almost all of Malone's financing," said Lorraine Spurge, who worked with Milken at Drexel Burnham. "TCI was a junk credit company."

TCI had little choice but to borrow money at high interest rates early on, but Malone and Magness were wary of the hostile takeovers fueled by junk bonds. So they did something that gave them total control and helped transform TCI from a struggling bit player into a monster of the midway. The innovation was supervoting shares—a plan cooked up in a bar, according to Sparkman.

Most public companies, including the 30 blue-chip stocks that make up the Dow Jones Industrial Average, are corporate democracies, with one-share, one-vote rules. Of course, company insiders, wealthy investors and mutual funds own many of those shares and gain decisive influence.

Supervoting shares allow insiders to maintain control without owning a majority of the stock. In 1979, TCI shareholders agreed to create a Class B stock, with 10 times the voting power of each share of Class A common stock. The reason given was to avert a hostile takeover. Over time, Magness and Malone ended up with most of the B shares, which gave them extraordinary power. TCI was a public company but run like a private one. Magness and Malone could access the public stock, debt and bond markets, but answered mainly to themselves. This gave them the freedom to build their business as they saw fit. Many family-owned media companies did the same thing.

Yet, the supervoting structure insulated TCI from outside accountability. The company had no independent board members and became known for stock deals that benefited insiders. An investor buying TCI shares really bought Malone's vision of the cable wire as the nerve center of home entertainment. He was more than happy to sell that vision to anyone who would listen. Even better if they had money.

A cable analyst named Jack Myers pitched Malone on a new venture during a 1983 snowstorm that blanketed Denver and TCI headquarters. Stripping off his soaked shoes and socks, Malone propped his bare feet on the desk and started selling. "In the first 10 minutes he told us why he wasn't going to fund us," Myers recalled. "And in the next three hours and 50 minutes, he told me his vision for the cable industry. The next day I went and bought the stock."

Malone had an appetite for scale economics, debt financing and doing deals by the dozen. A cable industry lobbyist named Robert Johnson departed with $500,000 from TCI in 1979 to help launch Black Entertainment Television, which he vowed would increase cable penetration in the inner cities. It did and TCI ended up with 35 percent ownership of the network. "Bob seemed like the kind of guy you could invest some money with and he wouldn't embarrass you," Malone said years later. "When we thought about starting this, there was always the concern that this kind of channel could become radical and you wouldn't want your name associated with it."

TCI flourished in the 1980s, spending nearly $3 billion in stock, debt and cash to gobble up 150 smaller cable operators, representing a majority of those that came on the market. In 1980, TCI was the nation's third largest cable company, with less than one million subscribers. By 1996, it was by far the largest, with 14 million.

The wager was backed by cable's surging popularity and by use of an accounting tool known as cash flow valuation. Many companies are valued on quarterly earnings, with net income paid out as dividends to shareholders or reinvested. Electric and phone utility stocks are valued this way and typically attract conservative investors.

Cable was much like a utility, but the companies rarely showed net income or paid dividends. They were valued instead on their ability to service the huge debt they took on to build and maintain cable systems. The key measure became earnings before interest, taxes, depreciation and amortization. This was known as EBITDA, or operating cash flow.

Malone and his financial team at TCI used cash flow as a lever to borrow billions of dollars for acquisitions and construction. For every dollar in cash flow, many times that amount could be borrowed. This required a predictable revenue stream, for which a monopoly cable system was a great guarantee. In fact, the faster cable grew its cash flows, the more money could be borrowed. The danger to this debt pyramid is rate regulation and competition. Either one could dry up cash flows and cause the pyramid to crack and perhaps crash.

The beauty of the debt pyramid comes from rising asset and stock market valuations, not the bottom line. Rather than pay corporate income taxes or distribute taxable dividends, TCI typically reported net losses and focused on building an empire.

"When you really start showing earnings, it means that you've run out of things to build and invest in," Malone said. "It's not conventional wisdom. Probably 5 percent of Wall Street thinks this way, 95 percent still plays the quarterly earnings game."

Supervoting stock control and cash flow valuations in the cable industry were, in large part, perfected by Malone, the financial engineer. Few people understood what he was up to, but the ones who bought into his long-term view became fabulously wealthy.

The cable gang bought, sold and traded cable properties at multiples of cash flow, called private market values, that were typically higher than the value placed on the properties by the stock market. The big payday would come when outsiders stepped in to pay what the cable gang said the systems were worth—and paid a premium for the supervoting stock.

The big payday would come, in other words, with the Greater Fool. This theory states that it is not foolish to pay $10 for anything at all, if someone else will buy it for $12. It is a wry description of capitalism. It all depends on the perceived value and sales pitch.

In the 1990s, Internet companies such as Yahoo!, Netscape and Excite had little debt, little cash flow—very little at all but a cool computer screen and an engine behind it that looked like the future to some investors. These companies raised hundreds of millions of dollars from stock sales and were valued at billions of dollars.

It all depends on the pitch.

In his industry, Malone had perfect pitch. He possessed a mathematician's conceptual skills, an engineer's training, an investment banker's facility with money, a flimflam man's patter and a contingency lawyer's hunger at the bargaining table.

His negotiating skills became legendary. One technique was to go back, or have his underlings go back, for a little more after a deal was struck, but not yet signed. At that point, it would be easier for the other side to give in than to walk away. Malone was also keen on the hedge-and-leverage game. He would rather own 20 percent of five interconnected, or even competitive, businesses than own 100 percent of one. Diversity made his power systemic without requiring that he actually run the businesses, which held little interest for him.

"What I really love is strategy," he said in 1998. "I hate running things. I hate getting subpoenas. I hate giving depositions. I hate getting called before a city council—I hate all that. And I hate having 35,000 employees who need to be patted on the back. Just give me a corner somewhere where I can just sit there and scheme, you know. Come up with creative structural ideas, technical or financial—that's me, that's my personality."

Companies and people are chess pieces, to be moved or sacrificed as logic dictates. Malone would be happiest by himself, moving the knights and rooks and pawns of his media empire, testing his theories on the world. Watching Malone play the game, it is tempting to call him a master chess player.

He is, in fact, a chess machine, like IBM's Deep Blue.

Disney chief Michael Eisner once recalled encountering Malone at the 1994 annual Allen & Co. summer conference for media moguls and their families in Sun Valley, Idaho. Eisner was so awed by Malone's brainpower that he shied away from him.

Many have had a similar experience with Malone. It is the visceral and unsettling sensation of being outwitted without quite knowing how. Thus a real reason for Malone's resemblance to the enigmatic and intimidating Darth Vader.

Of course, if someone wanted to play rough, Malone could do that too. TCI became a veteran of combat with broadcasters, movie studios, phone companies, programmers, satellite companies, the press, the Federal Communications Commission, the Federal Trade Commission, the U.S. Justice Department, the Internal Revenue Service, Congress and public officials ranging from the White House down to the mayor of Vail. Malone spent his business life flexing mind and muscle. He acknowledged this at times with a mixture of honesty, humor and condescension.

"To call us a giant, in the context of who we have to get up in the morning and compete with, is silly," he said in 1993. "If you want an analogy, I kind of think of TCI as the first generation of mammals in the world of dinosaurs. We can run around quickly, scurrying around and eating their eggs. And they're too dim-witted to stomp on us. If they ever do land on us, it's all over. But they're having a hard time hitting us."

WITH THE CABLE pipe in place, Malone viewed programming primarily as a commodity to buy and sell. He rarely let his personal likes and dislikes influence business decisions. "The garbage wouldn't be on," he once said, "if people didn't want to watch it and pay for it."

In 1984, he did send a telegram to Frank Biondi, then the chief executive of Home Box Office. Malone objected to an Italian movie, *Black Emmanuelle*, showing on HBO's sister network, Cinemax, because of its sexual content. Malone threatened to reconsider plans to expand Cinemax on TCI systems.

"We just kind of laughed at it," Biondi recalled.

Malone claimed to be a "documentary freak" who watched PBS, CNN and CNBC. He described C-SPAN as "great comedy" and prophesied that Turner's Cartoon Network, with its colorful animation crossing cultural borders, would become "the most valuable media property in the world." Yet, Malone disparaged one particular cartoon, the profane *South Park* on Comedy Central. He distanced himself from his company's decision to carry it. "Little kids using foul language," said Malone at an evening cocktail party at TCI in 1998, shaking his head. "If my guys had any guts, it wouldn't be on. But my guys don't have any guts."

Politically, Malone described himself as "a social libertarian and a fiscal conservative." He professed admiration for Ross Perot's independent run for the presidency and for conservative radio show host Rush Limbaugh, whom Malone once huddled with in New York. During the 1996 election cycle, TCI launched a weekly show titled *Race for the Presidency*, which gave free airtime to the major candidates. Consisting mostly of canned campaign ads and speeches, the show drew few viewers. Malone also launched a libertarian-slanted public affairs cable show on TCI systems called *Damn Right*, which broached subjects such as ending income taxation. TCI's deployment of *Damn Right* sparked controversy when *The 90's Channel*, a news program with a leftist bent based in Boulder, Colorado, was squeezed off seven TCI systems in 1996 after a massive hike in leased access rates controlled by TCI. However, when TCI's budget problems hit hard in late 1996, *Damn Right* also went off the air.

Malone did not let ideology, even his own, get in the way of business.

———

EVEN AS IT GREW powerful throughout the 1980s, Tele-Communications, Inc. remained virtually unknown, its name so generically corporate that it sounded odd to cite it without the "Inc." They operated in flyover country, Cowtown USA, far from the bright lights and media coverage of New York and Los Angeles. In TCI's hometown of Denver, the big business stories of the decade were the busted oil economy, and later, the savings and loan debacles.

For years, TCI's headquarters was a nondescript, one-story building, sometimes mistaken as a warehouse for set-top boxes in the Denver Tech Center, a sprawling business park south of the city.

There were no drivers and no company cars. Malone answered his own phone.

The company flourished without fully formed arms of the modern corporation: human resources, marketing and public relations departments. When an employee was hired or fired, that was human resources. When a new customer ordered cable, that was marketing. When Malone said, "We refuse to get raped by the programmers," that was public relations.

The chess machine focused on growth through economies of scale. By acquiring and maintaining more and more cable systems directly and through partnerships, he could command price discounts from programmers and increase the company's leverage in the industry. Malone's minions pursued the field part of this mission without too much agonizing. "Citizens in a community don't care at all who wins the franchise," said David Schultz, a TCI official, in 1980. "They just want cable. It's a behind-the-scenes political process."

People wanted cable. TCI had it. Case closed.

TCI, however, would face a costly comeuppance in Jefferson City, Missouri. The capital of the "Show Me" state endured TCI cable service outages, random picture quality and poorly trained field staff. When the company's franchise came up for renewal in 1981, the city hired a consultant named Elmer Smalling to evaluate bids from other companies.

Paul Alden was TCI's national director of franchising. He worked under J. C. Sparkman. Alden traveled to Jefferson City to take care of business. Alden told Jefferson City's mayor that TCI would turn off the cable unless its franchise was renewed. He threatened to destroy Smalling. "I was horrified by the sleaziness of everything," said Smalling in a 1998 interview. "It was a B movie."

In 1982, the Jefferson City council voted to award the new franchise to a TCI competitor, Central Telecommunications, which could either build a new cable system or, potentially, buy the existing one if TCI would sell.

TCI claimed a "First Amendment right" to remain in Jefferson City, with or without a franchise. The company withheld $60,000 in franchise fees due the city and announced it would rather have its system "rot on the pole" than sell to a competitor at any price. TCI's system manager told residents of a senior citizens' home, where antennas could not pick up TV signals, that TCI would cut off the cable and they would be without television for two years while a new cable system was built by a competitor.

After private negotiations between TCI and city officials, the mayor vetoed the ordinance awarding the franchise to Central. That left the franchise to TCI, which paid the withheld franchise fees. Central sued TCI for antitrust violations and tortious business interference. Central won a jury award of $10.8 million, plus $25 million in punitive damages.

TCI's actions, according to the trial judge in the case, were "nothing short of commercial blackmail." Alden, described by TCI as a rogue employee, was dismissed from the company. As late as 1986, Malone unsuccessfully pursued an appeal of the jury award to the U.S. Supreme Court. Then TCI paid up. TCI also remained the cable operator in Jefferson City.

"They never took the case itself seriously," said Stephen Long, the attorney who represented Central Telecommunications against TCI. "They run roughshod over little people and they think they don't have to answer for it."

TCI became much more circumspect after the Jefferson City case, but remained a study in contrasts. Its gunslinger ways played badly in the provinces, but worked miracles within the cable industry. Programmers and speculators poured through the doors at TCI, pitching deals, ideas and technologies. They wanted money or access to the cable pipe. Malone wanted, and got, a piece of all their businesses. He had a real-world laboratory to test his theories of debt financing, scale economics and programming investments.

John Hendricks, a former educational TV consultant to universities, founded the Discovery Channel in 1985, driven by the idea of documentary programming. The company was so broke his staff couldn't afford to send overnight mail packages. Several cable companies bought in to the channel. Malone took the lead and ended up owning 49 percent of Hendricks's company, Discovery Communications.

Then came the Turner bailout. Eager to expand his programming assets, Turner bought MGM's film library and other assets from Kirk Kerkorian in 1986, funded with $1 billion in junk debt raised by Michael Milken. Turner's company nearly tanked. Led by broker Bill Daniels, eight cable companies came to the rescue, with TCI and Time Warner ending up with 21 percent each of Turner Broadcasting System. Turner remained atop his company, but with two new bosses in the form of Malone and Time Warner's Gerald Levin, both of whom kept Turner on a tight leash.

Meanwhile, TCI trolled the high seas, eating smaller fish and spitting out the bones.

In 1988, TCI bought Tempo Enterprises of Tulsa, Oklahoma, for $100 million in TCI stock. Tempo's assets included the satellite service that delivered Turner's Atlanta Superstation, the Tempo cable channel that TCI later sold to NBC for $20 million, and a key satellite TV license.

As the sale closed, Magness told Tempo owner Ed Taylor, "Ed, you're the seventy-fifth company we have purchased and none of the executives have stayed with us."

TCI liked it that way, lean and mean. They were a tight-knit group of deal makers in Denver with a business plan long on generating cash flow and short on plant investment. Malone was the alpha dog. If you followed him, and didn't mind too much how he got there, you could get rich. Career secretaries at TCI were said to be millionaires through a pension plan funded with company stock. Observers were alternately amazed and aghast at the company's rise.

A *Fortune* story touted TCI's stock as one of the best performers of the modern era. One share bought at a low of $1 in 1974 was worth $913 in 1989.

A 1992 story in *The Wall Street Journal* took the opposite tack, tracing TCI's hardball franchise tactics and the acquisition by

Magness and Malone of TCI's Class B shares through a series of Utah cable system transactions so mystifyingly complex that probably only Malone knew the truth—and he wasn't talking.

———

THE ENORMOUS CHANGES at TCI had a parallel in the life of Bob Magness. In September of 1985, his wife Betsy died of a heart attack while they were on a horse-buying trip in Poland. The couple had been together for nearly 40 years, raised two sons and built a cable company from the ground up. She was the first, last and only woman to serve on TCI's board of directors.

Their passion was raising Arabian horses. During a trip to a horse farm in Phoenix, Bob and Betsy Magness had met a woman named Sharon Costello, who managed the facility. After Betsy Magness's death, Bob and Sharon became reacquainted at a horse auction and talked for seven hours. She was two decades younger than Magness, but they grew close and she eventually took over management of the Magness Arabian operations in Colorado and California.

They began living together in December of 1986, and married in 1989 after signing a prenuptial agreement. Her net wealth at the time was $508,000. His was $481 million.

———

MALONE, TOO, HAD become rich. He owned a 59-foot sailboat, a Mercedes that once belonged to Gary Cooper and a vacation home on 200 acres overlooking Boothbay Harbor, Maine. In the late 1990s, Malone and his wife acquired 14,000 acres of land in southeast Colorado. Malone handled some of the transactions himself, going door to door, introducing himself and asking if the landowners wanted to sell.

For all of his wealth and growing influence in the cable industry, Malone was not particularly status conscious or concerned with appearances. His personal taste ran to off-the-rack suits and Holiday Inn–style office furniture. Shuffling down a hallway by himself after a business meeting, or stopping to pour himself a cup of coffee,

Malone looked like what he was: a middle-aged engineer, who just happened to be one of the most powerful media executives in the country. With access to a cable pipe serving six million cable homes by the end of the 1980s, Malone began investing in more of the programming to flow through it.

His reach was ecumenical. In 1990, TCI invested $45 million in Pat Robertson's Family Channel. This allowed Robertson to separate the channel from his Christian Broadcasting Network and create a new public company called International Family Entertainment. TCI owned a minority stake, but the first right to buy should Robertson sell.

TCI spun off its programming assets and some cable ownership into a company called Liberty Media Corp. in 1991. By that time, Liberty held stakes in a group of regional sports networks and programmers, including Black Entertainment Television, Discovery Communications, the Home Shopping Network, Turner Broadcasting and many others. Malone was Liberty's chairman. The company was headed by Peter Barton, a street-smart and wisecracking executive who had joined Malone at TCI in 1982.

Swapping much of his TCI stock for Liberty's, Malone and those who followed him, including Barton and mutual fund manager Gordon Crawford, made fortunes when Liberty's stock went through the roof. By the end of the 1990s, Malone's stake in Liberty helped make him a billionaire.

—

IN 1991, A TCI EXECUTIVE named John Sie split off to create Encore Media, a programming company that created cable movie channels to compete with HBO and Showtime. Malone had hired Sie in 1972 at Jerrold Electronics. Sie was a native of Shanghai whose diplomat father moved the family to Taiwan in 1949 to escape the Communist takeover of the mainland. The family immigrated to the United States the following year. Sie had a bachelor's degree in electrical engineering and a master's degree in electrophysics from the Polytechnic Institute of Brooklyn. After working at Showtime, he joined TCI in 1984 and served in a range of strategic, policy and programming roles.

Since Encore was 90 percent owned by Liberty Media, it had an inside line to the TCI cable pipe. In test markets, the Encore movie channel was initially offered free of charge. But subscribers then had to notify TCI if they *didn't* want the channel, or be billed several dollars a month. This so-called negative option ran afoul of attorneys general in several states and was quickly dropped. Sie remained unapologetic. "It was a good idea that turned into a political nightmare," he said.

Malone's maneuvers and TCI's aggressive moves in the cable and programming arenas attracted ongoing scrutiny on Capitol Hill, and at the U.S. Justice Department, which created a wall map charting TCI's cross-investments and influence. Over the course of two decades, Malone and the cable gang had built a powerful union of pipe and programming that dominated the markets it served.

"There's nothing wrong with having a monopoly," Malone once said. "A patent is a monopoly. The real question is how do you behave in relation to your competitors?"

Malone was an Information Age version of John D. Rockefeller, except Rockefeller did it on a much larger scale. He became history's first billionaire in 1913 only after the Standard Oil Trust was broken up into companies that became Exxon, Mobil, Chevron and others.

The modern-day revenues of Exxon alone dwarfed those of the entire cable industry. Yet, the similarities are striking between the pipeline businesses of oil and cable and the methods of Rockefeller and Malone. The hunt for oil set off frenzied exploration, with boom-and-bust cycles moderated by Rockefeller's economies of scale. Cable's wildcat early days and franchising frenzy were moderated, to a great degree, by Malone's economies of scale.

Standard Oil owned one-third of the nation's oil wells at its peak, while refining 90 percent of the country's oil. TCI controlled 28 percent of the nation's cable homes at its peak, either directly or through partnerships. Its Liberty Media programmers, the equivalent of oil flowing through the pipe, supplied every cable home in the nation.

Standard Oil colluded with railroads to win special transport rates, demanded bulk discounts on crude oil purchases and used its leverage to crush competitors. TCI demanded "most favored nation" rates from its vendors, required volume discounts on programming purchases and used its leverage to undermine competitors.

"Rockefeller envisioned a new industrial order based on monopolies and trusts instead of competitive struggle," according to biographer Ron Chernow. Malone professed that "there is more profit in peace than war" and created interlocking investments that resembled a trust.

A devout Baptist, Rockefeller believed that God guided his every move, regardless of the tactics needed to heed that guidance. Malone's ethos was New England rationalist: destined to succeed by following the ironclad dictates of reason.

Standard Oil was broken up in 1911 after the trust-busting efforts of President Theodore Roosevelt, and by the Sherman Act of 1890, which became one of the major federal guidelines in merger reviews and antitrust enforcement. The crackdown came only after the trust, and Rockefeller himself, were dissected and vilified by muckraking journalist Ida Minerva Tarbell. In his rise to power, Rockefeller had crushed small oil refiners in western Pennsylvania, including Tarbell's father, Franklin. Ida Tarbell wrote a 19-part history of the Standard Oil Company for *McClure's Magazine* beginning in 1902. Widely read at the time, the series combined blow-by-blow recitations of Rockefeller's business deals with needling insights into the man himself. In a section titled "The Oil War of 1872," she wrote:

> Thus, on one hand there was an exaggerated sense of personal independence, on the other a firm belief in combination. ... Mr. Rockefeller was "good." There was no more faithful Baptist in Cleveland than he [but] religious emotion and sentiments of charity, propriety and self-denial seem to have taken the place in him of notions of justice and regard for the rights of others.

Unhampered, then, by any ethical consideration, undismayed by the clamor in the Oil Regions, believing firmly as ever that relief for the disorders in the oil business lay in combining and controlling the entire refining interest, this man of vast patience and foresight took up his work.

Malone faced his own muckrakers in the 1980s and 1990s. They weren't journalists for the most part, though journalists and the mass media amplified claims by consumer advocates and corporate rivals that Malone had become too powerful for the public good.

Malone's muck was raked instead by a Tennessee tornado that consisted of an ambitious congressman from Carthage named Albert Gore Jr., and a satellite TV dealer from Oak Ridge named Charlie Ergen who started out on the road to Deathstar in a ditch.

FIVE

CUTTING CABLE

C HARLIE ERGEN WHEELED the Mark V Lincoln Continental down a dark stretch of Colorado highway past midnight. Jim DeFranco, his business partner and poker-playing buddy, rode shotgun. Rumbling along behind the car was a trailer with a 10-foot satellite dish on its side, chained down, like a concave fiberglass sail. It was December of 1980. Ergen and DeFranco, both 27 years old, were about to crash into the satellite TV business.

Their first customer was a wealthy rancher in Pagosa Springs, a town in southwest Colorado, 25 miles north of the New Mexico border. The rancher asked Ergen and DeFranco to buy a TV set and bring it down with the satellite dish. They bought the TV set in Denver, but DeFranco's Lincoln was rear-ended by a car during the shopping trip. That ate up a few hours with police reports and frayed nerves. It was nightfall by the time Ergen and DeFranco got on the road with the satellite dish in tow.

Outside the town of Walsenburg, high winds caught the rigid hollow of the dish, torqued the trailer and broke it loose from the car's tow bar. Half their company's inventory lay sideways in the passing lane of Interstate 25. In shock, Ergen parked the car alongside the highway's sunken median strip. He said later it was the second-worst experience of his life, the first being his father's death.

DeFranco got behind the wheel. He backed the car up, then used the front bumper to slowly plow the crippled trailer and battered satellite dish into the shallow of the median strip. Then they drove silently into town for help.

This was not the bright start envisioned by the young hustlers.

Ergen and DeFranco had met three years before in Dallas. Ergen was a financial analyst at Frito-Lay. DeFranco was a wholesale liquor salesman whose previous gig was selling Kirby vacuum cleaners door-to-door. Ergen wanted to buy a color TV and heard that DeFranco had one for sale. DeFranco wanted $350 for it. It was a monster Zenith in a walnut cabinet with a stereo built in—a real piece of furniture.

DeFranco turned the set on. The screen filled with a black-and-white picture.

"Where's the color?" asked Ergen.

"Oh, it just needs to warm up," said DeFranco, fiddling with the control knobs, telling Ergen how great the picture would be, how the color would pop in anytime now.

Who's this guy trying to kid? thought Ergen.

—

CHARLIE ERGEN GREW up in Oak Ridge, Tennessee, a Levittown of the Atomic Age. Along with Hanford, Washington, and Los Alamos, New Mexico, Oak Ridge was built and run by the U.S. government for the sole purpose of creating the Manhattan Project atom bomb for use in World War II. The government had seized farmland to build three facilities separated by ridges of the Great Smoky Mountains, with easy access to the hydroelectric power of the Tennessee Valley Authority. There was K-25, a gaseous diffusion plant. There was Y-12, an electromagnetic plant. And there was X-10, a plutonium plant.

The research operations were surrounded by a barbed-wire fence with armed guards in towers. The town itself was built 10 miles up-valley from Y-12. At its peak in 1945, the population was 75,000. Every man, woman and child had to wear an identification badge.

William K. Ergen worked at Oak Ridge as a nuclear physicist and specialist on reactor safety. He was a native of Austria who earned his Ph.D. from the University of Vienna. He immigrated to the United States prior to World War II as part of the European brain drain. He met his wife, Viola, in Minnesota, and they moved to Tennessee in the early 1940s. Most of William Ergen's work was classified, but at least one aspect of his research later became well-known: the "China Syndrome," describing what would happen if a reactor core melted down, creating heat so intense it would burn a hole toward the center of the earth.

Oak Ridge settled into domesticity after the war. The families were young and the breadwinners generally well paid. The Ergens raised five children, the fourth named Charles William Ergen.

On the night of October 4, 1957, William Ergen took four-year-old Charlie for a walk. They and others in town—and around the world—watched as *Sputnik I* tumbled around Earth every 96 minutes. It made a light streak across the sky, emitting a Morse code message that touted the superiority of Soviet technology and signaled the start of the space race.

Charlie Ergen was less interested in the basketball-sized satellite than in basketball itself. He played almost every day with the neighborhood kids, made his high school team, and the freshman team at the University of Tennessee at Knoxville. A small forward at just under six feet tall, Ergen said the game taught him to catch fire inside to overcome adversity. "When you bring a level of intensity and emotion, you can perform better than people think you can."

When his father died of a heart attack in 1971, Charlie—just graduated from high school—began paying his own way through college. He clerked nights at the athletic dorm, printed checks at a bank and sold fraternity sportswear. His own fraternity, Phi Gamma Delta, was where he learned to play poker, sometimes winning stereo systems and textbooks from his house brothers in lieu of cash. "Staying sober while playing poker was definitely an edge," he said.

From a huge public university, Ergen joined a small class at Wake Forest University, where he earned his MBA. He took his CPA exam and worked as a traveling auditor of textile plants for Collins &

Aikman Corp., a textile company based in Charlotte, North Carolina.

Ergen moved to Dallas to take a financial analyst's job at Frito-Lay, where he soon grew bored. He hung out with Frito-Lay and Braniff Airline employees at a bar called Daddy's Money, which eased the pain of the workweek with three-for-one happy hour drinks. It was there that he met, and later began dating, Cantey McAdam, a Braniff flight attendant whose Irish first name was contracted to Candy.

One day in 1978, he told Candy he was "retiring" from Frito-Lay to live off his savings and $15,000 in stock saved for him by his parents. "I thought he was crazy," Candy recalled. "My mother thought he was crazy. I said to him, 'You're only 25. How can you retire?' He said, 'I can't work for someone who's not as smart as me.'"

Ergen became a freelance bum. He scoured *The Wall Street Journal* for business ideas in the morning, hit tennis balls in the afternoon, then played poker at night with DeFranco and others at Ergen's apartment complex. He invested in the stock market and began traveling overseas, hopping flights with Candy, who got 90 percent discounts on airline travel and hotel stays. They flew to Hong Kong, Tokyo, Sydney and Paris.

Ergen fancied himself a trend spotter. He saw people wearing cowboy boots on the streets of Paris and connected it to the booming country western nightlife he and DeFranco witnessed at a Dallas club named Cowboy. "I learned as an auditor that the way to research a company wasn't to talk to the executives, but to the other people in the company," said Ergen. "I went down to talk to the guys at Tony Lama in El Paso. They said, 'Man, we're shipping twice as many cowboy boots this year as we were last year.'" Ergen and DeFranco bought Tony Lama stock. Nothing happened. When the John Travolta movie *Urban Cowboy* came out in 1980, they thought they'd cash in. Still nothing.

They sold the stock, missing out on a rally in Tony Lama stock a few weeks later.

Ergen was always hunting for an angle to avoid working for

someone else. Flying on Braniff's half-empty planes while airline deregulation loomed, Ergen bet in the stock market that the price of Braniff shares would fall over time. "I was flying on Braniff and shorting their stock," he recalled fondly. Ergen made some money, while Braniff continued a nosedive on its way to bankruptcy in 1982.

When he and Candy flew to Seoul, Ergen visited the 24-hour casino at the Sheraton Walker Hill Hotel. He saw young Americans at the blackjack table, alternating large and small bets, winning big. The blackjack players ended up at a party thrown by the Braniff flight crew at the hotel. Ergen asked about their betting strategy and they told him about card counting: Players assign a value to each card dealt, keep a mental tally, then bet heavily when the odds favor the player over the house. When Ergen returned to Dallas he bought a book titled *Playing Blackjack as a Business.* He called De-Franco, who bought his own blackjack book. Together, they studied the cheat sheets. Because casinos rely on fixed odds and gamblers' greed to rake in profits, they frown on players counting cards to beat the house.

"Sir, I'm going to have to ask you not to play 21," a security guard at a Lake Tahoe casino informed DeFranco, whose blunder was lip-synching the numbers as the cards were dealt.

"I figured I made about $5 an hour when I wouldn't get kicked out," said Ergen. "I realized I wouldn't make a living doing that. Plus, it was really boring. For every card, there was a right and a wrong decision. You couldn't have a more boring profession."

Ergen, DeFranco and Candy soon embarked on a much bigger gamble, the satellite TV business. It would eventually bring them riches and into the orbit of two master game players, Rupert Murdoch and John Malone.

Candy moved to Phoenix to get her MBA from Thunderbird International School of Management. Ergen moved with her. De-Franco stayed in Dallas. One Sunday in the early fall of 1980, he drove to a friend's house to watch a Dallas Cowboys football game on television. On the way, he saw a van parked by the side of the road with a beige trailer carrying a color-coordinated, 10-foot satellite dish. DeFranco stopped and knocked on the door of the van. The man inside was watching football games via satellite from all over the country. The van and dish were a mobile display of a sales franchise the man had bought from a company called Starview in St. Petersburg, Florida. DeFranco took some notes and called Ergen the next day. They decided to fly down to Florida and check out Starview.

Plenty of franchise territory was left, given that only 5,000 satellite dishes—at prices ranging from $10,000 to $36,500 apiece—were sold at retail in the entire country in 1980.

Ergen, DeFranco and Candy took a chance. They pooled $60,000 in savings and bought a satellite dish franchise for Colorado, Utah and Wyoming. Denver was the biggest city in the Rocky Mountain region, so they moved there, setting up shop in a 900-square-foot storefront on Broadway in Littleton, just south of Denver. They placed a big satellite dish in the parking lot as a visual advertisement. Candy flew up from Phoenix on weekends. DeFranco, living on food stamps, bought furniture on his credit card for the group house they rented and shared. Prominent in the living room was the monster Zenith TV DeFranco had failed to sell.

Ergen, returning from the library one day, told DeFranco he had a great name for their new company: Sputnik.

"Nobody will buy from us," DeFranco objected. "They'll think we're a bunch of Commies."

They settled instead on Echosphere, a combination of two names. "Echo I" was the name of an experimental NASA balloon satellite. "Sphere" referred to the parabolic shape of the satellite dishes.

The Echosphere owners opened their doors on Thanksgiving weekend 1980, and waited.

And waited.

After days playing backgammon in the deserted store, the entrepreneurs realized that an urban area was an unlikely place to sell big satellite dishes. Their first customer was the Pagosa Springs rancher.

The sale was actually made by the man in the van in Dallas, but Echosphere would do the installation and split the $3,000 commission with the salesman.

—

THE PROBLEM WAS, one of their two dishes was mangled along Interstate 25. Ergen and DeFranco hired a tow truck to transport the trailer and dish into Walsenburg, while they drove west through Alamosa and South Fork to Pagosa Springs. They pulled up to the ranch just after dawn—unshaven, sweaty and exhausted. The TV set and set-top box were in the backseat. Sticking out of the trunk was an eight-foot metal pole to support the missing dish. Taking one look at Ergen and DeFranco, the ranch foreman told them he'd take care of the concrete pad and support pole and sent them on their way. The Echosphere duo collected their damaged inventory in Walsenburg and drove back to Denver with a strange story to tell.

Before Christmas, they returned to the ranch with the spare dish in tow and claimed their first paying customer. To celebrate, they hacked down a pine tree by the side of the road on the way back and tied it down on the trailer. Returning home in triumph to Candy, they discovered that their pirated Christmas tree had fallen off.

Business improved from that shaky start. They took out ads in rural newspapers, selling satellite dishes for up to $15,000, with half up front, plus $600 for installation and $1 a mile for any delivery over 50 miles. If a buyer supplied room and board for an overnight stay, Ergen and DeFranco waived the $1-a-mile charge. Deep in the mountains, where cable TV lines did not run, people sometimes flagged down Ergen and DeFranco on the road to ask about the dish on the trailer. On a working vacation, Ergen and Candy trolled the streets of tony Aspen and sold a dish to a sheik for $18,000.

They cut ties to Starview and began buying equipment and making industry contacts on their own. The Echosphere trio traveled to trade shows in places like Omaha, Nebraska, where the hotel meeting rooms were filled with card tables covered with butcher paper. The vendors, mostly gearheads and video nuts, displayed their wares and traded tips. "There were no experts," said Ergen. "No one knew any more about it than I did. And we were too stupid and too naive to know that we couldn't do it ourselves."

Someone who knew more than Ergen was Bob Luly, whom Ergen met after Luly spoke at a satellite show in Houston in 1981. Ergen was by no means sold on the satellite business as a way to make money. He had discovered the feel of eel in Seoul and was seriously considering an import business based on selling eelskin wallets, briefcases and other goods.

"The main thing he wanted to be was a distributor," said Luly. "He didn't care if it was shoes, magazines, eelskin wallets or satellite dishes. The margins were high on satellite dishes, so he liked that. He never took any money out of the company and didn't want anybody else to, either. He drove around in this rusted-out car and put all the money back into the company, like a slot machine."

Ergen's focus on distribution and a willingness to bet it all gave his company an uncanny durability amid tempests of technology, regulation and competition. "We were having a good time," said Candy. "We thought, If it works, great. If it doesn't, well, we can always get a job somewhere else." Two years after the Thanksgiving weekend that Echosphere opened for business, Candy and Charlie were married. Ergen chose that date because they were already scheduled to get three days off work. Two years after that, they had the first of their five children.

They moved the company to the Inverness Business Park, about two miles south of TCI's headquarters. Echosphere took a stab at the nascent small-dish satellite TV business in 1985 by forming a company called Antares Satellite Corp. They owned 30 percent. A. B. Hirschfeld Press, a Denver-based printing company, owned 10 percent. The remaining majority stake was held by United Cable Television Corp., headed by Gene Schneider, who started out with Bill Daniels in cable's Wild West days.

Antares didn't get very far. "United Cable decided that they didn't want to proceed," said Ergen.

Schneider later sold United Cable to TCI and kept its international assets to form United International Holdings. Antares Satellite Corp., a company in name only, thereafter listed its president as John Malone on incorporation papers.

Echosphere prospered. It shifted from retail to wholesale distribution and cultivated a network of independent dealers. Ergen controlled its business from top to bottom. When original equipment was hard to come by, he bought a manufacturer named Houston Tracker Systems. When it came time to sell programming, he created a sales arm called Satellite Source. And when some dealers wanted to buy on credit, he started Echo Acceptance Corp.

Echosphere set up offices in London, Bombay, Singapore and Beijing. Ergen claimed his staff had the ability to speak 30 languages. So important was the overseas market that Echosphere began using the trade name "EchoStar" because "sphere" was awkward for Asians to pronounce. By the end of the 1980s, the privately-held company was on its way to annual sales of $200 million. Then came the battles with the big boys.

AS HBO HAD PROVEN in 1975, satellites were a powerful way to deliver national programming to the growing patchwork of cable companies—provided that they invested $100,000 or so for 33-foot satellite dishes. "TCI was so poor that our bank covenants wouldn't allow us to have that much money in one lump," Malone said. "So a bunch of us put our money together, bought one, and put it in Minot, North Dakota."

The efforts of backyard dish builders like Taylor Howard, salesmen like Ergen and the mass production of satellite dishes drove the retail market and lowered prices. By the mid-1980s, there were more than one million dish owners, with 90,000 people a month signing up.

A key advocate was Al Gore.

The son of a famed U.S. senator by the same name, Gore had a privileged upbringing. His birth was announced on the front page of *The Tennessean* on April 1, 1948. The family had a 250-acre farm just east of Carthage, near the Caney Fork River. Gore's early life was split between the farm and the Fairfax Hotel in Washington, where his parents lived for much of the time while the elder Gore served in Congress. He was an author of the Interstate Highway Act and fought utility rate hikes for his largely rural constituency.

For nine years, Al Gore Jr. attended the exclusive St. Albans School for Boys in Washington. He was captain of the football team his senior year, the liberal party leader in government class and nicknamed "Gorf." He met his future wife, Mary Elizabeth "Tipper" Aitcheson, in 1965 at the school's graduation party.

Gore graduated from Harvard University with a degree in government and spent another year at Boston University. He opposed the Vietnam War, but said his conscience obliged him to enlist. He joined the U.S. Army in 1970 and served seven months in Vietnam as a military information officer. His father was a vehement opponent of the war. The stand cost him his Senate seat in the 1970 election.

On his return to the States, Gore Jr. took graduate classes at Vanderbilt's divinity school and worked as a reporter at *The Tennessean*. His travels took him to The Farm, a commune in Summertown, southwest of Nashville. One of The Farm's projects was a homemade satellite TV system. Intrigued, Gore later enlisted two residents of The Farm, Mark Long and Douglas Stephenson, to install a satellite dish at his parents' farm in Carthage, where there was no cable TV service.

Gore wrote a range of stories at the newspaper, but thrived on exposing municipal corruption. He enrolled in Vanderbilt's law school in 1974 and later became an editorial writer at *The Tennessean*. In 1976, before completing his graduate work, he ran for a vacant congressional seat in Tennessee's fourth district, a post once held by his father.

Junior, 28 years old, was in the House. Gore's 15 years in Congress were marked by an intense work ethic, a fascination with public policy and a talent for playing hopscotch with a wide range of issues. He had a taste for technology, including satellite communications, CB radios and computers. "He was excited when he got a watch for Christmas that did all this goofy stuff," said Mike Kopp, Gore's press secretary from 1981 to 1988.

Gore was an early proponent of government use of electronic mail. He began using the term "information superhighway," a modern echo of his father's role in helping legislate the interstate highway system. Gore chaired the House Science and Technology Subcommittee and lobbied House Speaker Tip O'Neill to have House proceedings televised on the fledgling C-SPAN, a network funded by the cable industry.

During C-SPAN's inaugural coverage of House proceedings in 1979, Gore was the first House member to speak from the floor. He soon got his mug on camera as often as possible, as did another ambitious southern hambone, Republican Newt Gingrich of Georgia. Gore also beat the bushes back home, averaging 200 town meetings a year during his eight years in the House. In his travels, he saw hundreds of big satellite dishes sprouting up in rural areas.

In 1983, Mark Long, the man who had helped install a satellite dish at the Gore Sr. farm, asked Gore to speak at the satellite TV industry's convention at the Opryland Hotel in Nashville. He also asked Gore to help with national legislation to allow big-dish owners to legally receive programming via satellite. By this time, there were more than one million home satellite dishes in use. A popular model was eight feet in diameter, priced at about $3,500. To lure customers, some vendors advertised "free HBO" and "100 channels," which programmers and cable companies considered promotion of piracy.

Roy Neel, Gore's chief of staff, was wary of Gore taking on HBO and cable programmers on the piracy issue, according to Long. But Gore worked the seam. In the 1984 Cable Act, which lifted cable rate regulation, there was a provision written by Gore that affirmed the right of home satellite viewers to pluck programming from the sky. The act also gave programmers the authority to scramble their signals. Scrambling involves coding satellite signals so that only viewers with proper equipment can see the TV pictures clearly.

"This bill removes the legal cloud which has made many potential customers wary of satellite," said Gore, who won a U.S. Senate seat that year.

But the war clouds were just forming.

Just as broadcasters welcomed cable companies until they became a competitive threat, the cable gang embraced satellite technology up to the point where hundreds of thousands of people owned dishes. To protect their signals, the cable programmers pursued satellite encryption in the early 1980s.

In satellite TV, as was true with the printing press, movies, cable and home video—and would be with the Internet—a market innovator was a service with a committed audience: pornography. Porn channels on satellite scrambled their signals and required dish owners to pay several hundred dollars for a decoder box, plus $100 or more a year for programming. There were niche markets even in that. The Fantasy Unrestricted Network, or FUN, boasted XXX movies, but drew the line at R-rated slasher flicks. "It's one thing to show people nude," explained Chuck Dawson, Fantasy's president. "It's another thing to hack them up while they are nude." A competitor to FUN, the Pleasure Channel, countered with soft-core fare. "I won't show penetration into any orifice," vowed Norm Smith, the channel's president, "whether it be by penis, vibrator or banana."

The significance of the porn channels' signal scrambling and payment scheme was not lost on mainstream cable programmers, which claimed to be forgoing $10 million a month in revenues from satellite piracy. Public service networks, like C-SPAN, and ad-supported ones, like the Home Shopping Network, remained unscrambled. But the premium cable networks like HBO, CNN and ESPN began scrambling in 1986. Big-dish owners were eventually required to buy a decoder for $395, plus programming fees.

By this time, Taylor Howard was chairman of SPACE, the Society for Private and Commercial Earth Stations. He urged his industry to gain legitimacy and distance itself from the signal piracy that had helped fuel the boom in big dishes. "We suffer to some degree from wounds inflicted by a few piranhas and snakes in our own industry," he said in early 1986. "As a group they have done nearly irreparable harm to our industry by focusing the wrath of the cable and broadcast industries upon us."

The scrambling by major programmers caused big-dish sales to plummet, sending the industry into a tailspin and forcing hundreds of marginal vendors out of business. There was a grassroots revolt, particularly among dish owners in rural areas of Tennessee, Kentucky, Arizona and Kansas. Located beyond the range of broadcast TV and unwired for cable, the rural dish owners claimed the right to capture whatever flew through the air over their own property.

In North Carolina, "You've got to have a pickup truck, a gun rack, and a satellite dish on the side of your house," said Democratic representative Charlie Rose.

Scrambling "is one of the most emotional issues I've dealt with since coming to Congress," said Utah Republican representative Howard Nielson during a 1986 hearing. Outside the Capitol, satellite dealers set up trucks, one of which sported a hand-lettered sign, "Free My Airwave."

The power plays among cable companies, dish vendors, programmers and regulators led to some guerrilla tactics reminiscent of TCI's 1973 actions in Vail. In the early morning hours of April 27, 1986, HBO viewers on the East Coast settling in to watch *The Falcon and the Snowman* were treated instead to this message on their TV screens: "Goodevening HBO from Captain Midnight. $12.95/month? No Way! (Showtime/Movie Channel beware!)"

Stunned HBO engineers in Hauppage, Long Island, fought to overtake Captain Midnight's signal. They finally got the movie back on the air after a few minutes. Captain Midnight turned out to be John MacDougall, a 25-year-old employee of the Central Florida Teleport, an uplink facility that delivered satellite programming to cable operators. MacDougall had a job on the side selling big dishes, which wasn't going well due to the scrambling controversy. While on duty at the Teleport, MacDougall jammed the HBO signal with the Captain Midnight message. He later pleaded guilty and was given a year's probation, fined $5,000 and had his radio operator's license suspended.

While scrambling by cable programmers rocked the big-dish industry, Malone was already making his play. TCI created a big-dish leasing operation, with programming packages. Yet, he objected to programmers independently aligning with satellite services and bypassing cable ownership. When HBO was delivered directly by satellite to big-dish customers, the charge was $12.95 a month, with $5 of that rebated to the local cable operator to keep the peace. Showtime engaged in a similar practice. Antitrust complaints logically followed.

In September of 1986, Senator Gore co-sponsored a bill to allow third-party vendors to sell programming directly to home satellite viewers. One of the vendors interested was Amway, the direct marketer of household and beauty products. Gore's Democratic Senate colleague from Colorado, Tim Wirth, prevailed upon HBO, Showtime and the cable gang to end the rebates practice, undercutting the bill.

The Reagan administration also opposed the bill, as did Republican senator Barry Goldwater of Arizona, soon to retire. He helped defeat it by offering a comparison that showed dish owners in Arizona paid less for a comparable programming package than did cable subscribers.

Yet, there was no free market to test supply and demand. Who controlled access to programming? At what price? According to Gore, the answers were the cable gang and too much. At one hearing, Gore cited Pat Robertson as a victim of cable's strong-arm tactics in 1988:

> He was preaching the gospel, he was running for President and he was raising money hand over fist and he would seem to have every incentive to want as many people watching his service as possible. When he provided it in an unscrambled form to satellite dish owners, he was intimidated. He was threatened by the large multisystem cable operators and told, point blank, if you do not scramble, we

will take you off our large cable systems in the cities. And since most of his viewers were still in the cities, he yielded to intimidation. He was shaken down, in effect.

The story was provocative. Gore had obtained a letter sent from Robertson's office to a Canadian satellite subscriber describing the pressure. But the story, according to Pat Robertson's son, Tim, wasn't true. "We couldn't afford to be in business without encrypting the signal," said Tim Robertson, president of The Family Channel, in a 1999 interview. "I wrote to Senator Gore and asked him to please stop waving that letter around. It had been sent out by an assistant in my father's office to a Canadian viewer, but the facts were wrong. I followed it up in a meeting with Senator Gore and his chief of staff, Roy Neel, but they wouldn't listen. It was nothing but political grandstanding to become the champion of the home dish owner."

Said Neel, "That's bullshit. The letter states on its face. We didn't distort it, but reported it as it was."

During Senate Commerce Subcommittee hearings on satellite TV in the summer of 1987, Gore cited an unholy alliance between programmers and their cable company owners and patrons. He claimed that Ted Turner's proposal to package 20 cable channels for sale to satellite viewers was shelved after TCI and other cable companies bailed out Turner's company that year.

In fact, the pressure on Turner had come two years earlier with a threatened lawsuit by TCI. Turner also faced difficulty getting copyright clearance to offer broadcast channels as part of the package.

While Gore bypassed inconvenient facts or ignored mitigating points of view in order to tell a better story, he did capture the spirit of the problem. "The marketplace for home satellite dish owners is clearly not working," he said. "It is not competitive and it will not be competitive until the interlocking cable programming companies release their stranglehold on a fledgling technology that has brought the communications revolution to millions of rural families."

One of the committee witnesses was fellow Tennessean Charlie

Ergen, who also portrayed satellite TV as a liberating force suppressed by the cable gang. "We were too good," Ergen testified. "We became a competitive threat to people who owned the old technology."

It was rabble-rousing stuff, televised live on C-SPAN. But the real pounding came when Ergen took on General Instrument Corp., the vendor of set-top boxes for TCI and other cable companies. In 1986, GI bought MA/COM, a company that made the VideoCipher II, the industry-standard satellite TV decoder. Priced at $395, the decoder allowed big-dish owners to purchase the same programming as cable subscribers. Ergen's company sold the decoder and even integrated it into its set-top box. Yet, GI could not meet the pent-up demand for the VideoCipher, nor would they license Echosphere as a subsidiary manufacturer.

With GI executive Larry Dunham sitting nearby, Ergen lit into one of his primary suppliers. "The initials GI in our industry stand for Gouging Instruments," said Ergen, pointing out GI's 35 percent profit margin to Gore and other senators. Ergen also claimed that GI knowingly sold compromised decoders that allowed pirates to watch satellite programming for free, which "put honest satellite dealers out of business."

Ergen's stinging public attack raised his profile in the satellite TV industry. It also made him a target.

Eight months after the hearing, on April 1, 1988, Echosphere executive Steven Schaver sat in the company's Miami office, explaining to a new employee what a calm place it was to work. Just then, agents of the U.S. Customs Service, with guns drawn, raided the warehouse. Brandishing a search warrant, they backed a truck up to the loading dock and seized 7,600 VideoCipher II units. Similar Customs raids occurred around the same time at Echosphere facilities in Dallas, Texas, and Tempe, Arizona. A snowstorm in Denver delayed the agency's raid on Echosphere headquarters, which gave Ergen enough time to get his lawyers involved and prevent a complete shutdown of company operations.

Customs officials claimed that Echosphere illegally sold GI decoders to Mexico and the Bahamas in violation of an Arms Export Control Act provision. The act claimed that components of the decoder could be used for military purposes. General Instrument, eager to export its equipment, opposed the restriction. But they also helped the federal government crack down on violators.

Customs officials alleged that Echosphere offices kept two sets of books and shipped the decoders through Texas border towns. The source of this tip was an anonymous informant that turned out to be a GI employee, according to David Drucker, Echosphere's attorney at the time. Over the next 20 months, an Echosphere subsidiary and several employees, including Schaver, faced federal charges. Echosphere settled the case by paying a fine of $800,000 in the form of satellite dishes for schools in Colorado.

"We weren't buttoned up as a company with all the controls we should have had," Ergen said in 1998. "When they raided us, it almost put us out of business. We had every government agency going through every record. It made me realize that competitors can use federal agencies against you. And that if you're going to be out front, you're going to be a target. It shows the power."

———

AL GORE ALSO SHOWED his power. In 1988, the Satellite Home Viewer Act passed Congress and was signed into law. It allowed home dish owners to receive out-of-market signals from NBC, CBS, ABC, Fox and PBS if they lived in "white areas" where local broadcast signals could not be received by a conventional off-air antenna. It was a breakthrough for the satellite TV industry. But it led to a breakdown over the next decade as broadcasters alleged in court and in Congress that satellite TV firms sold the signals far beyond white areas.

Gore pursued a head-spinning number of issues beyond cable and satellite TV in Congress, from infant formula to organ transplants, from arms control to the environment. He tried to translate his legislative agenda into a mandate for the presidency in 1988. He said all the right words to his Democratic constituency, but came off like a mannequin.

"Where's your passion?" the Reverend Jesse Jackson, a fellow presidential candidate, challenged Gore when they met on the road.

"We never operated from a political strategy, we just chased opportunities," said Mike Kopp, Gore's former press secretary. "If you look at a list of his hearings, it was all over the map. It was like a buffet he was sampling. We tried to do that on a presidential level and it was a different game. He never had great political instincts. He did have a keen sense for opportunities that might make headlines."

As Gore's presidential hopes faded, one of those opportunities fell into his lap. In 1988, a fund manager and former ABC executive named I. Martin Pompadur ran a company called MultiVision that overpaid for cable systems that served 29 communities in west Tennessee. Overpaying for cable properties was not uncommon after rate regulation was lifted in January of 1987. The problem was the remedy, which typically involved jacking up rates. Some of these situations were purposeful. Speculators bought high, raised rates and tried to sell even higher. There were ominous calls from Congress on the cable industry to end this trafficking or face the consequences.

Some cable systems were just poorly managed.

MultiVision raised rates as high as 46 percent, causing howls among subscribers. Gore had a hot pocketbook issue and he toted it, holding town meetings and earning ink as the voice of the consumer. In the summer of 1989, MultiVision rolled back its rates with the intention of putting "old tensions firmly behind them," according to a letter sent to Gore from a company executive. Nice try.

"It's good news to see the company respond to local concerns and complaints," said Gore. "But it doesn't remove the underlying problems created because in cable television there's no competition and no protection from massive rate hikes."

The cable gang had a good thing going with rate deregulation and they blew it. "The cable industry has been praying for more than two decades to be free of government interference in the setting of basic rates," Bill Daniels wrote to his colleagues in April of 1986. "Nine months from now, that dream will become a reality. The question is, will it end up being a boon to the cable industry, or a nightmare?"

By December of 1986, Daniels had answered his own question. "Some operator is going to overplay his hand when it comes to deregulation and end up generating a lot of bad publicity and backlash for the entire industry. I don't know who it will be, but when you give people the freedom they've asked for, someone, somewhere inevitably screws up."

By 1989, rate hike controversies prompted the introduction of 11 separate bills in Congress to re-regulate the industry. Gore was at the forefront. "I would want the whole country to know that it all began in West Tennessee and that we have a better cable system because of you," he told a meeting of the region's mayors and town officials in August of 1989. "The cable industry is powerful in Washington and [re-regulation] will depend largely on public support for passage."

To keep up the populist pressure, Gore needed a villain. I. Martin Pompadur, though he had a name worthy of a Charles Dickens character, wouldn't do. Not big enough. Gore found one from central casting in John Malone. By 1989, Malone ran a company with six million subscribers and stakes in Turner Broadcasting, Black Entertainment Television, American Movie Classics, the Discovery Channel, sports networks and other programmers—all of which greatly worried the broadcasters. "Meet the man who makes the networks tremble," wrote *Business Week*.

Neel, Gore's chief of staff at the time, recalled his first impressions of Malone in a 1999 interview: "cold, business-like, politically arrogant, but brilliant, visionary and tough as nails."

And when he came to Washington?

"A fish out of water."

Cable's stunning ascendance in the 1980s had provoked the sclerotic broadcasters, the Hollywood studio cabal and the rate-regulated phone companies, all of which had long-standing political and campaign funding ties in Washington. Cable had none of that.

"I knew we had lost the battle on that front when I was involved in government relations for TCI," said John Sie, one of Malone's top lieutenants. "I said, 'Geez, what is the size of US West, a regional phone company, what is the size of their Washington office?' They must have had 70 people total. Imagine how many lunches they can take people out on, multiplied by 300 days a year. Their life depends on lobbying at the state and federal level. And cable is like the frontier entrepreneurs. We have two people in the Washington office and say, 'Hey, we've arrived.' Forget about cable's internal problems. Cable was just a total neophyte when it came to perception."

Cable was a blessing in disguise for Hollywood, creating lucrative new outlets for movies on pay-per-view and premium channels, including ones launched by Sie in the early 1990s, Encore and Starz!. But in the late 1980s, the studios feared that the cable gang would eat its way up and down the entertainment food chain. Jack Valenti, the silver-haired and silver-tongued chief of the Motion Picture Association of America, worried in 1988 that nothing less than the American way of life was at risk unless the backs of the cable gang were broken by regulation. "To convey to a handful of corporate owners a kind of absolute power over television, the most pervasive social and political influence in the land, is to invite a most subtle kind of infection whose vanity and potential arrogance could have a wicked effect on the body politic," he said.

Gore, a less flowery demagogue, studied the economics of the cable, programming and nascent satellite TV industries. He came to the accurate conclusion that cable was a monopoly. Yet, Gore was so rigid that he could not stand to have his views challenged and demanded the acquiescence of those who appeared before him at congressional hearings. "The witnesses that got off the easiest were ones that showed some humility," Kopp recalled. "He didn't like anyone to imply that they were one step ahead of him. He had a huge ego when it came to that. The witnesses that came in with some arrogance were in trouble."

Enter Malone. Testifying before Congress, or even dealing with the Washington bureaucracy, was repellent to Malone. He could not see the correlation between his concerns—namely, building cable systems, generating cash flow, making investors rich—and what he viewed as the misguided focus of the federal government: taxing, spending and making stupid rules.

In business affairs, Malone said, "Government should be mainly a cheerleader." In Gore's view, government was the referee, sometimes the coach and maybe even the quarterback. Their conflict was bred in the bone. Gore's father was a New Deal Democrat who idolized Franklin Roosevelt. Malone's father abhorred Roosevelt and had just one association listed in his 1978 obituary: membership in a group called Citizens of Milford Battling Against Taxes.

When Malone appeared before the Senate Commerce Subcommittee on Communications in November of 1989, Gore lay in wait. He rattled off the total cable rate increases in Tennessee in the previous three years: Nashville, 113 percent; Chattanooga, 115 percent; Murfreesboro, 117 percent.

"Some just say, 'Well, that's the marketplace at work,'" said Gore. "Well, it's a monopoly. And the federal government has come in and told the local governments, 'Hands off. Let them fleece the consumers as much as they possibly can.'"

Gore then set the hook in Malone's mouth.

"TCI, for one—and we'll hear from its CEO today—is obviously hell-bent toward total domination of the market as it buys up not only more and more cable systems, but more and more major programming services, and even movie studios. Earlier this week, in another Senate committee there was considerable concern about the Super Bowl possibly moving to pay-per-view cable in the near future. I think that they are focusing on the wrong danger. I think a real concern might be whether or not TCI will own the NFL outright."

When Malone's turn came, he cited cable's diversity of programming, the billions of dollars invested and the increasing subscriber counts. Acknowledging the pressure to re-regulate the industry,

Malone even donned sheep's clothing for the senators sitting in judgment. "You know, we really are not bad guys," he said. "We really are not trying to build an empire. We really are willing to listen to the directions of the government."

When Gore and Malone squared off, it was a classic confrontation between a pompous politician and a ruthless businessman. Gore came on like Ida Tarbell going after Rockefeller, or Robert Kennedy taking on union boss Jimmy Hoffa in the 1950s organized crime hearings.

> GORE: Do you think that your size and power have made your company arrogant and heavy-handed?
>
> MALONE: I don't beat my wife, either. No, I honestly do not believe so. I think that the actions that we have taken have been almost entirely, well, they have been entirely really trying to serve as honest purchasing agents for our subscribers.
>
> GORE: Have you ever threatened a locality in connection with a franchise renewal?
>
> MALONE: Well I don't know what constitutes a threat. I think we have indicated that if we don't get renewed, under federal law we cannot continue to serve.
>
> GORE: Well, let me read to you from court testimony what I think constitutes a threat. One of the top people in your franchise renewal department, according to court testimony, "threatened a consultant hired by Jefferson City, Missouri, saying, 'We know where you live, where your office is, and who you owe money to. We are having your house watched. We are going to use this information to destroy you. You made a big mistake messing with TCI. We are the largest cable company around. We are going to see that you are ruined professionally.'" Are you familiar with that case?
>
> MALONE: I lost an awful lot of sleep over that one, Senator.
>
> GORE: You were fined $25 million in punitive damages?
>
> MALONE: That's correct, in a civil antitrust case.
>
> GORE: It was $10.8 million in actual damages. Judgment affirmed on appeal.

MALONE: That's correct.

GORE: Is that an aberration?

MALONE: That's an aberration.

That exchange marked the beginning of the end of cable's experiment with deregulation. Anti-cable sentiment swelled into a tidal wave over the next three years, helped along by Gore's rhetorical flourishes. "Darth Vader" and "Cable Cosa Nostra" became media tags for Malone and the cable gang. Malone did not defend himself publicly, telling associates that he preferred not to get in a pissing match with a skunk.

The cable chiefs, used to being masters of their own domains, could not form a unified front to strike a compromise with Congress to re-regulate their industry. Glenn Jones, the founder and chief of Jones Intercable, once described attempts at industry unity on any number of issues as "trying to get Samoan warlords in a kickline."

TCI, Time Warner, Cox, Continental, Comcast and Cablevision all had different views of what to do. Malone reportedly tanked one deal in 1990, preferring to take his chances with Republican president George Bush, who was likely to veto any re-regulation bill passed by a Democratic-controlled Congress. In any event, a mishmash of regulation would not outwit the chess machine. "John has a favorite saying—you tell us what the rules are and we'll tell you what we look like and how we are going to operate," said Bernard Schotters, TCI's vice president of finance, in 1990.

But a mishmash is exactly what happened.

Rather than come to grips with what rules might support a competitive marketplace for the long term, Congress took a middle path strewn with perks, pitfalls and political posturing. The 1992 Cable Act mandated that programming owned by cable companies be made available to competitors at a fair price, if the programming was sent up on satellite. Small-dish satellite TV companies like GM Hughes, EchoStar and Advanced Communications—none of which had launched service—fought hard for that provision, because they would not be able to compete effectively without it.

Yet, there was a broader issue that Congress did not face. A bold proposal would have been to split ownership of programming from distribution completely, an act for which there was precedent. In the 1940s, Twentieth Century Fox, Warner Bros. and other movie studios also owned most of the nation's movie theaters. They dictated what movies would play where and for how long, effectively strangling competition from independent theater owners. A U.S. Supreme Court ruling in 1949 forced the studios to divest their theater ownership, a prohibition that remained in force. The antitrust restriction was such that the Premiere channel, an exclusive pay-TV service owned by the studios and featuring their movies, was blocked by the federal government in 1980.

A second issue Congress ducked was whether to keep cable companies from the nascent direct-to-home satellite TV business. They could have required the cable gang to divest Primestar, its satellite TV service, which was then under a U.S. Justice Department antitrust investigation. The failure to address this issue would loom large in coming years.

A key provision of the 1992 Cable Act further entrenched the power of the broadcasters. It allowed them to charge cable companies for programming they otherwise delivered free through the air. Rather than pay, the cable gang engaged in side deals to carry cable networks owned by broadcasters.

The centerpiece of the 1992 Cable Act was intended to punish the cable gang for their customer service misdeeds, rate hikes and monopolist mind-set: Congress would order the FCC to cut cable rates, purportedly to levels that would exist if effective competition existed. This made for great sound bites, but muddled public policy.

Democrats and many Republicans piled onto this bill in the summer of 1992. How could they go wrong bashing cable? Senate co-sponsors of the bill included Gore, who by that time was running for vice president, and Republican John Ashcroft of Missouri.

The vote was 280–128 in the House and 74–25 in the Senate, majorities big enough to override a presidential veto.

The national political context was the upcoming presidential election between Bush and Democratic candidate Bill Clinton. In early October, Bush vetoed the cable act, claiming it would not lead to lower prices for consumers. Up to that point, he had a perfect string of 35 vetoes without an override by Congress.

"Cable consumers are getting ripped off and George Bush is giving the cable monopolies permission to do it," said Gore on the campaign trail. Bush lost in more ways than one. The House voted to override the veto on October 5, 1992, by a vote of 308 to 114. Some Democratic members chanted, "Four more weeks! Four more weeks!" an allusion to a Bush reelection slogan of "Four more years." Democratic and Republican legislators from only one state unanimously supported the Bush veto: Colorado.

In the Senate, Bush came within one vote of winning. When it became clear that it wouldn't happen, several Republicans were released from their commitments. The veto override drew 74 senators, including 24 Republicans. It was a remarkable bipartisan coalition a month before a presidential election.

"Bush's veto was anti-consumer and we just drove it down his throat," said Roy Neel, Gore's chief of staff, who became president of the United States Telephone Association in 1994. As for the cable gang, he said, "We had them by the corporate personals and weren't going to let go."

Though many factors played into the 1992 Cable Act, it will forever be known by some as "the John Malone Retribution Act." A prominent cable TV executive who worked closely with Malone in that era said this:

> John was off the reservation. TCI was being abusive to their customers, to the towns they worked in. John is the smartest technical guy in the business. I've worked with all sorts of guys in broadcasting and entertainment. John's better than those guys. But what's the price you pay for that? Intellectual arrogance. He recognizes

politicians for what they are and he says it. In terms of public sensitivity, he should not be the industry leader, the front man. If you're going to be a statesman for the industry, you've got to be more disciplined.

Even so, government-imposed rate regulation was a gimmick. It did crimp cable revenues by 17 percent, but also led to unintended consequences. Out went nonrevenue-producing C-SPAN on many cable systems; in came the home shopping channels. More channels were taken out of the basic cable package and offered à la carte. Smaller cable companies suffered the brunt of across-the-board rate cuts.

Malone remained defiant. "We'll continue to diversify away from the regulated government-attacked core," he said as the new rules took hold in 1994. "And meanwhile, we'll continue to slug it out in the trenches in the domestic cable business, recognizing that the government's got to kill a lot of smaller cable companies before they can really hurt us much."

The cable gang was forced to fill out reams of compliance paperwork, like juvenile delinquents serving detention. The FCC hired several hundred new bureaucrats to shuffle the paper. The rate cuts initially took several billion dollars out of the cable rate base. But by the late 1990s, with a new deregulatory scheme, cable rate hikes averaged 8 percent a year, more than three times the general inflation rate. Some in Congress expressed amazement at this and suggested regulating the industry once again.

Yet, only competition and innovation would truly force the cable gang to fly right. That pressure would come from above, from the very satellite technology that had helped cable flourish in the first place.

SIX

THE RAT ZAPPER

T HE PROPHET OF WAR between satellite TV and cable was Arthur C. Clarke. The famed science fiction writer, co-creator of *2001: A Space Odyssey*, grew up in the 1920s seaport of Minehead, England. He was fascinated by the Atlantic Ocean and the feeling of bodily weightlessness in the water. He doodled spaceships during school and haunted the local Woolworth's to buy pulp magazines such as *Amazing Stories* and *Astounding Stories of Super-Science.*

At home, Clarke and his friends spied the surface of the moon with crude telescopes they crafted from lenses and cardboard tubes. He earned a few pence building and selling wireless radio sets. He also worked in the Bishops Lydeard post office, where his grandfather had been postmaster.

At age 18, Clarke took a civil service exam and qualified for an entry-level auditor's job in London. There, he joined the British Interplanetary Society, a group of dandified space and science fiction enthusiasts who ate fish-and-chips in a Bloomsbury flat and dreamed up ways to send a manned expedition to the moon.

Clarke published his science fiction writing until World War II. He enlisted in the Royal Air Force in 1941 and became a flight lieutenant, training airmen to maintain radar gear used to land aircraft in spotty weather. A self-schooled expert in ballistics and space flight, Clarke wrote a paper grandly titled "The Future of World Communications." After getting clearance from the Air Ministry, he submitted the 3,000-word article with four diagrams to a journal called *Wireless World*. It was published in October of 1945 under the more pedestrian title "Extraterrestrial Relays."

His original title was better. The article was a remarkable blueprint for the global satellite communications industry. It came 12 years before the Soviets launched *Sputnik I* and three years before Ed Parsons hooked up his cable TV connection in Astoria, Oregon.

Clarke envisioned manned space stations launched 22,300 miles above the Earth's equator, a zone later to be known as the Clarke Belt. Solar-powered, the satellites would orbit at the same speed the Earth rotates, five miles per second. Viewed from Earth, the satellites would appear to be motionless, like fixed moons that "neither rise nor set," Clarke wrote. Communications signals would bounce from an uplink on Earth to the satellites, then back down to "small parabolas perhaps a foot in diameter" that "once adjusted need never be touched again."

Imagine a powerful searchlight high enough above Earth that its rays covered a continent or two. This is something of what Clark envisioned. Three satellites, evenly spaced and relaying signals, could provide global coverage. This was Clarke's elegant solution to barriers faced on the ground. TV signals sent from radio towers eventually fade away. Unless towers dotted the landscape, Clark wrote, only "highly populated communities will be able to have television services" while the "problem of trans-oceanic services remains insoluble."

The cable cowboys eventually lassoed those tower signals and ran them through miles and miles of coaxial wire to hundreds, and then millions, of homes in the United States. Satellites literally transcended ground-based communication by beaming voice, video and data from above. "No communication development which can be imagined will render the chain of stations obsolete and since it fills what will be an urgent need, its economic value will be enormous," wrote Clarke.

The economic value to Clarke was slight. He sold the article to *Wireless World* for $40 and never bothered to patent the idea of communications satellites, which became a multibillion-dollar international industry. Clarke, who moved to Sri Lanka in later years, was deemed the "Godfather of Satellite Communications" for his 1945 article. But he demurred, "I suspect that my early disclosure may have advanced the cause of space communications by approximately 15 minutes. Or perhaps 20."

It would take two decades before rocket propulsion, satellite technology and government regulation caught up with Clarke's dynamic vision. But once the idea took hold, satellite signals wrapped the world. In the blink of an eye, satellites sent—and made no distinction among—billion-dollar bank transfers, battlefield video and a phone call from Texas to Tangier.

Military ballistics research and Cold War passion drove the development.

Less than four months after *Sputnik I* went up in 1957, the U.S. Army launched a 30-pound unmanned satellite into orbit, earning a triple-decker, front-page headline in *The New York Times*. In 1961, President Kennedy made his famous plea to a joint session of Congress that America land a man on the moon by the end of the decade. Less celebrated was Kennedy's request that the United States also help build a global satellite communications network. Even farther from public view was J.C.R. Licklider, chief of computer research at the Defense Advanced Research Projects Agency, who wrote memos about a "Galactic Network" to transmit data around the world.

While NASA pursued the moon shot, and Licklider the Internet, Congress in 1962 created the Communications Satellite Corp., known as Comsat, for commercial deployment of satellites. It may have been the last time that U.S. policy kept pace with the breakneck speed of communications technology. The first satellite TV transmission, an eight-minute broadcast from France to the United States, occurred via AT&T's Telstar satellite that same year.

Though it was a government-created monopoly, Comsat's funding came from a $200 million stock sale, the largest initial public offering at that time since Ford Motor Co. in 1956. Half the shares were snapped up by communications giants like AT&T, ITT, GTE and RCA. Half were purchased by the public.

On April 6, 1965, Comsat launched its first satellite, nicknamed *Early Bird*, from Cape Canaveral, Florida. The launchpad was 150 miles away from a site at which Jules Verne, the nineteenth-century science fiction writer, imagined a 900-foot-long cannon buried in the earth with enough explosive force to launch a rocket to the moon. Among those watching *Early Bird's* launch on closed circuit video from Comsat's Washington, D.C., offices was Verne's visionary descendant, Arthur Clarke.

Comsat and its international counterpart, Intelsat, helped convert military-based swords—the science of rockets and spy satellites—into commercial plowshares used by phone, TV and other communications companies to send signals worldwide. Yet, when Comsat and other corporate titans tried to deploy that space-age technology to deliver satellite TV direct to homes in the United States, they crashed. The list of failed entrants and missed opportunities is a Who's Who of business, including Sears, Alcoa, NEC, Western Union, RCA, Murdoch's News Corp., CBS, Gulf + Western and Prudential Insurance. Tens of millions of dollars were burned and at least two satellites ended up in the ocean.

The lure was easy to explain. The rise of cable TV indicated a lucrative hunger for channel choices. Cable counted 18 million subscribers by 1980, but that left 65 million homes without, one-third of which might never be reached with cable due to the expense of stringing wire into the American outback. Satellite was no discriminator by distance, geography or trench costs. Once the bird was in the sky and the uplink built, it was just a matter of flipping dishes underneath, like buckets to catch the rainwater.

The big satellite dishes satisfied some of the demand among ranchers and hobbyists. Yet, even when dishes shrunk to a diameter of eight feet and a price of $1,000, they were too bulky and expensive to capture a mass market. A bonanza awaited those who could deliver satellite TV to much smaller dishes at a price competitive with cable. That service became known as Direct Broadcast Satellite.

Some said DBS stood for Don't Be Stupid.

"Direct by satellite is never going to be competitive with cable," John Malone said in 1981. "Even if they go to the small dish. The investment per channel is so much lower in cable. You have all kinds of maintenance costs with DBS and a whole satellite system up there you have to pay for. It's just not competitive."

He was right at the time. But as DBS grew viable over the next two decades, the chess machine made a series of moves with just one intent: to delay DBS for as long as possible while trying to gain control of it himself. Of all the strategic games Malone played, this was the longest running, the most filled with reversals and the most critical. If Malone misplayed this game, his financial engineering of the cable industry might unravel. Even when Rockefeller was building the Standard Oil Trust, he didn't face the kind of killer competitor that satellite TV represented to cable.

By the early 1990s, Malone referred to DBS by another, more evocative, name: the rat zapper. The rats were subscribers. If DBS zapped too many of them, the cable gang's carefully constructed debt pyramid could begin to crumble and the river of monthly subscriber payments would divert to the promised land of satellite TV.

Getting DBS off the ground first required international agreements to allocate orbital locations, the satellite parking spots in the sky. The United States and other countries met in Geneva for a five-week International Telecommunications Union conference in 1983 to divvy up the airspace. As with the United Nations, the one country, one vote rule led to geopolitical power plays. Ecuador and Colombia claimed, but failed to win, sovereignty over orbital slots above their countries. In an earlier meeting, Israel was nearly tossed out of the union in a bloc vote led by Arab countries.

The United States was awarded the eight DBS slots it coveted. Yet, only three of the slots allowed a launched satellite to "see" the entire continental United States. These three slots were designated by their locations at 101, 110 and 119 degrees longitude over the central and western half of the country. The nine degrees of separation is key. It means that satellites can deliver clear signals to dishes as small as 18 inches in diameter. Satellites spaced two degrees apart, called medium power, require much larger dish sizes on the ground to avoid signal interference.

Medium power is a Chevy. DBS is a Ferrari.

Controlling any one of the three key slots became a survival strategy as DBS evolved. Of the three, the middle slot at 110 is the prize. Owning 110, plus either one of the adjoining slots, would allow a satellite TV company to potentially deliver hundreds of channels and other services to a single 18-inch dish.

Throughout the 1990s, Rupert Murdoch, John Malone and Charlie Ergen waged a stealthy battle for 110. By turns, they bought, threatened, undermined, partnered and sued to gain control of it. Each time one of them closed in, one or both of the others—or some other entity, like the federal government—threw a blindside hit and the billion-dollar game of cutthroat began anew. The motives of Murdoch, Malone and Ergen were as varied as their personalities. With 110, Murdoch would complete a master plan to dominate satellite TV worldwide. In Malone's hands, 110 would ensure cable's dominance, perhaps by keeping it unused. Ergen needed 110 to advance his aerial assault against the cable gang: 110 was the ultimate rat zapper.

Before that battle commenced, many contenders stormed the DBS shores in the 1980s. They ended up bleeding in the celestial surf. Comsat was first. Even before the DBS slots were allocated by the U.S. government, Comsat's subsidiary, Satellite Television Corp., proposed spending $683 million to launch up to three satellites, the first one by 1985. Three channels, one each of movies, sports and culture, would be offered initially. The subscription price would be $25 a month, plus $100 for the dish. Comsat projected two million subscribers within three years of launch, even though the market for a three-channel service appeared slim and the space-based agency lacked some earthly skills.

"Retailing and TV programming are two areas that Comsat knows absolutely nothing about," Wilbur Pritchard, a former company executive, said at the time. Joe Charyk, Comsat's president, had earlier acknowledged the risk: "If nobody wants to subscribe, obviously it would be an economic disaster."

It took four years and $200 million for the disaster to reveal itself. Comsat ordered two satellites from GE Astro, then attempted and failed to recruit other companies as partners, including CBS and Gulf + Western. A 1983 report from the Yankee Group, a technology research firm in Boston, urged Comsat to scale back before DBS became its "corporate Vietnam." The Comsat board scrapped its DBS plans in December 1984 without launching a satellite or signing up a single customer. Comsat's failure was crippling for the company, for the fledgling DBS industry and particularly for an entrepreneur named Robert W. Johnson.

Johnson was a U.S. Navy veteran, a professor at the University of Detroit and the founder of a market research firm. He worked briefly in 1979 at Pat Robertson's Christian Broadcasting Network, which delivered *The 700 Club* and other shows to cable systems via satellite. Like Robertson, Johnson sensed a big untapped audience for religious and family programming. Johnson created Dominion Video Satellite in 1980 and applied to the government for a DBS license while raising money and seeking equipment vendors.

"The cable industry and broadcast TV hoped the whole idea would die," said Johnson. "There was a huge financial cost associated with staying the course."

The broadcasters bashed DBS just as they had cable. "In light of its potentially disruptive social, political and economic impact on local broadcasting, we do have serious questions about its desirability," the National Association of Broadcasters wrote to the FCC in 1980.

Johnson pressed on, lining up an impressive cast of vendors by late 1984. Hughes Electronics, not yet owned by General Motors, would build two satellites. General Dynamics would launch them. NEC, the Japanese electronics firm, would make and sell the decoder boxes. A partnership between NEC and Alcoa, the aluminum company, would make and sell the dishes. GE Industrial Credit would give loan guarantees to the vendors. Total initial cost to launch and market the service: $200 million.

Johnson and executives from the participating companies met at Hughes' offices in El Segundo, California, to finalize plans. Midway through the meeting, a Hughes executive poked his head in the room to relay the news that Comsat was bailing out of DBS. The unspoken admonition filled the room: Don't Be Stupid. If Comsat, which pioneered the commercial satellite communications industry, couldn't make it work, why bother?

"We went and had lunch and couldn't regroup," said Johnson. "It threw cold water on the idea. It killed it for everybody." Another 12 years would pass before Johnson's DBS service, called Sky Angel, was launched—on Charlie Ergen's EchoStar.

Other DBS entrants took Comsat's departure as an omen. Western Union and RCA gave back their licenses to the federal government. Comsat sold its two unused satellites to foreign entities to recoup some of its investment. Completing the snakebitten scenario, both rocket launches failed on launch and the satellites ended up as sea junk. Before bowing out of DBS, Comsat considered, but rejected, a last-ditch merger with a company called United Satellite Communications Inc. (USCI), which had its five-channel satellite service up and running in the Midwest and Northeast by November 1983.

It was the original Deathstar.

Rather than launch its own DBS service, which took many months of planning and hundreds of millions of dollars, USCI spent $1 million a month to lease space on a medium-powered Canadian satellite already in orbit. USCI was backed by Prudential Insurance, which put up $45 million for 40 percent of the company. Twenty-five percent was held by Francesco Galesi, a New York real estate developer. Holding 14 percent was General Instrument, the cable equipment vendor that intended to manufacture 3,000 dishes a day to meet demand.

One hurdle was attaching a three-foot-diameter dish to the roof of a house without high winds ripping it off. "Isn't it amazing that with all this space-age technology our most difficult problem now is attaching the dish?" mused GI chief Frank Hickey, the man who beat out Malone for GI's top job in the 1970s. A bigger problem was the likelihood of USCI's financial roof caving in. About 11,000 customers signed up in cities including Indianapolis, Chicago and Washington, D.C. But it was a tough sell, with a $300 up-front equipment charge and $39.95 a month for programming. By early 1985, USCI was out of money and floundering.

One of the bottom feeders was Malone, who saw a glint of the future in the troubled start-up. "It had very limited channel capacity and didn't make economic sense," Malone said in 1998. "We used to laugh about it and call it the Deathstar, because of *Star Wars.*" The movie had resonance beyond that. In *Star Wars*, the Death Star is the flying fortress from which Darth Vader exercises his villainous powers. When Al Gore later pinned the name on Malone, some TCI employees bought black Darth Vader masks as a tribute, of sorts, to their chief.

Malone may have disparaged USCI's venture later on, but he tried to take control of it at the time. His motive was ingenious. TCI would turn USCI into an urban satellite service, delivering cable programming to city rooftops and apartment buildings—whether or not TCI had a cable franchise in those cities. This reflected Malone's overriding business philosophy, provocative to cable brethren, rivals and regulators alike: If there is going to be competition, it is better to *be* the competition. TCI and GI could not agree on terms, however, and the original Deathstar imploded and went out of business.

Meanwhile, the cable gang intended to remain the sole direct-to-home distributor of video programming in their regions, regardless of technology. GE Americom—a division of General Electric, which also owned NBC—had medium-powered satellites in orbit and formed a joint venture with HBO called Crimson Satellite Associates in 1985. It would deliver programming by satellite to cable companies. But Crimson bled red by 1988 because the cable gang refused to use it, fearing Crimson would eventually deliver

HBO and other programming by satellite direct to homes, bypassing the cable gateway.

Two years later, Malone gained control of a partial DBS license at the 119 slot when TCI bought Tempo. "One reason they wanted Tempo was to keep us the hell out of the DBS business," said Sel Kremer, a Tempo executive. Tempo's chief, Ed Taylor, worked at TCI for several years on DBS plans, including a potential alliance with GM Hughes Electronics. Taylor's business projections showed DBS taking a big chunk out of cable's core business. When Malone saw that, he backed away from launching DBS, said Taylor.

TCI faced another hurdle at the FCC. The Wireless Cable Association, two consumer groups and a DBS applicant objected to TCI getting into the DBS business at all, due to its strong-arm tactics in Jefferson City, Missouri, and subsequent $35.8 million fine. "TCI has a demonstrable track record of cutting off competition, both inside and outside the cable industry," read part of a petition to the FCC from a DBS company named Advanced Communications Corp. "To allow it to gain a foothold in the DBS industry would give TCI excessive power over other DBS operators and competing cable providers to the detriment of the public."

Malone bided his time, watching from the sidelines as one after another DBS entrant fell from the sky. Two of them belonged to Murdoch.

Murdoch was mainly known as a newspaper publisher in the United States in the early 1980s. As an Australian citizen, he faced restrictions on owning broadcast TV stations. Plus, cross-ownership rules blocked ownership of newspapers and TV stations in the same market. Eager to make his mark in American television, Murdoch gravitated to the new frontier, satellite.

"There are really no limitations on what one can broadcast over this kind of TV," he said in 1983, proposing what News Corp.'s annual report called "the first nationwide satellite-to-home broadcast network in the United States." This was three years before Fox TV.

Named Skyband, Murdoch's satellite service planned to sell dishes four to six feet in diameter, capable of receiving five channels of movies, sports and news at prices comparable with cable. With typical fanfare, Murdoch signed a five-year, $20 million contract to lease space on a private satellite launched in 1981 by the space shuttle *Columbia*, the world's first reusable spacecraft.

Murdoch discussed Skyband's potential with Daniel Ritchie, then the chairman of Westinghouse Broadcasting, which owned cable and broadcast properties. "I said, 'Rupert, you're going to get killed,'" Ritchie recalled. "He had no programming."

Murdoch also had limited channel capacity. And a dish size too big. Seven months after announcing Skyband, it was time to disband. Murdoch was stuck with a $600,000 computer billing system, an operations center in Secaucus, New Jersey, and a $12.7 million payment to get out of the satellite lease. As the proposed service slid from view, Skyband president Harvey Schein asked, "Do you know anybody who wants five 20-watt transponders on a good satellite?"

Murdoch remained undaunted. He pursued satellite TV in the United Kingdom, which was relatively unwired with cable and ripe for service. His Sky TV venture started slowly, offering four channels by 1989. It grew into a powerhouse. In the United States, Murdoch hurdled foreign ownership restrictions on broadcast stations by taking the oath of citizenship at the New York City federal courthouse in September of 1985. There was some payback in this, expressed by Murdoch during a tortuous legal battle over his ownership stake in Warner Communications, a company he coveted.

"If they think they can beat me by exploiting the fact I'm not a citizen, I'll become a fucking citizen and shove the deal straight up their noses," Murdoch said. He did, cashing out News Corp.'s investment in Warner and turning his attention to creating Fox TV, which became profitable by the end of the 1980s.

By that time, DBS appeared viable. The power of satellites had greatly improved, with a new technology promising to provide many more channels: digital compression. It grew out of frenzied research on high definition television, the technology that promised movie-like clarity and depth on TV screens. In the race between the United States and Japan to launch HDTV, engineers at General Instrument and other companies discovered digital compression.

Here's how it works. In a movie scene where a plane flies through the sky, very little may change for hundreds of frames except the position of the plane relative to the sky. By encoding the plane and sky digitally, much less information is required to display the same scene, as compared to analog waves, which constantly refresh the TV picture. In addition, the digital video stream is sent in bits of zeros and ones, like computer data, which provides super-sharp pictures and sound. Digital compression was critical to HDTV development, while causing seismic changes in the cable and satellite TV industries.

One analog cable channel could potentially carry 10 or more digital channels. Each DBS slot is licensed for 32 transponders, also called channels. Digital compression could potentially provide 10 or more channels on each one of those, or 320 total, like multiple warheads.

Rupert Murdoch was ready to go to war—sort of. Emboldened by his success with Sky in Britain, Murdoch gave DBS another shot in the United States. This time, he had partners: NBC, Cablevision and Hughes Electronics, the General Motors subsidiary that owned a DBS license, and was once part of the Howard Hughes empire.

At a February 1990 press conference, Murdoch and his Sky Cable partners proposed a 108-channel DBS service. They would sell flat-panel satellite dishes, 12 by 18 inches, for $300 apiece. News Corp., NBC and Cablevision were to put up $75 million initially to fund the $1 billion entry. The rest of the money would come from Hughes, plus bank and stock financing. Yet, Sky Cable faced the same problem that Ritchie pointed out to Murdoch years before: Where would the programming come from? The cable industry controlled it.

"There's no way that we, ourselves, could provide programs for

108 channels," Murdoch admitted at the press conference, adding that networks like HBO and ESPN would be "welcomed."

Cablevision chief Charles Dolan, the man who had founded what became HBO, described Sky Cable as "the future of cable television." Rather than compete with cable, he claimed, Sky Cable would become a satellite-delivered programming adjunct for subscribers, without the need to dig up streets and lay new cable lines.

The description was clever, but Sky Cable was compromised. Like Comsat before them, Murdoch, Dolan and NBC accurately saw the future of satellite TV, but lacked the nerve to seize it. Wary of offending fellow broadcasters, neither Murdoch's Fox nor NBC planned to put their own broadcast channels on Sky Cable. Meanwhile, Dolan could not pitch Sky Cable as a radical new competitor to cable, given that his company's core business was cable and cable programming. Sky Cable's mixed message undermined its credibility. "If the Sky Cable partners wanted to downplay the impact their proposed high-power DBS service would have on the video marketplace, they succeeded in spades. From one side of the country to the other, reaction ranged from skeptical to unconvinced," noted *Broadcasting* magazine.

"Dolan wanted it to be an adjunct to cable, Murdoch wanted to compete," said an executive involved in the negotiations. "NBC was on the fence and Hughes was passive. Everybody was glad that it tanked."

Beyond its internal conflicts, Sky Cable needed big bucks, which Murdoch, for one, was in no shape to provide.

By the end of 1990, News Corp. was nearly $8 billion in debt, including its $3.2 billion purchase of *TV Guide*, which Murdoch made without consulting his lawyer or investment banker. Murdoch's binge obliged him to pacify bankers from Zurich to Pittsburgh who had a claim on News Corp. through a web of loan syndication. The pirate's empire teetered on the edge.

One of News Corp.'s biggest drains was Sky TV in Britain. Though gaining subscribers, Sky lost about $3 million a week. It faced bruising competition from British Satellite Broadcasting, a mismanaged, but well-funded, competitor. Murdoch himself took over day-to-day operations of Sky and agreed, under pressure from lenders, to merge with his satellite rival in November of 1990, a deal that likely saved both companies from ruin.

By that time, Sky Cable in the United States was doomed. None of the three partners with Hughes had put up the promised $75 million. Both News Corp. and NBC expected that new government rules would guarantee satellite companies access to cable-controlled programming. But those rules would not come until the 1992 Cable Act. Tom Rogers, an NBC executive, claimed that the network's enthusiasm for DBS had "substantially cooled."

While NBC cooled, Malone plotted.

On the day Sky Cable was announced, Malone revealed plans for a satellite TV service backed by nine of the largest cable companies, which served 60 percent of the nation's cable subscribers. Named K Prime and later renamed Primestar, they planned to lease space on a medium-powered satellite owned by GE Americom, also a Primestar partner.

If DBS was the rat zapper, Primestar was the mousetrap. Its power and number of channels would be limited. But it gave the cable gang a relatively inexpensive inside lane to the rat zapper. Primestar was a "transitional technology" that would make the move to "true" DBS, said Robert Thomson, TCI's public policy chief, in 1990.

GE Americom had been searching high and low for partners to lease space on its medium-powered satellites. Murdoch was a potential buyer. So was Charlie Ergen before Malone stole his eyeballs. "I had a letter of intent all ready to go," Ergen said. "The head guy at GE Americom was coming to Denver to meet with me the day after the national cable show in Dallas. He met in Dallas with Malone, didn't come here, and ended up doing a deal with Primestar. That was the first time Malone out-maneuvered me."

It would not be the last.

SEVEN

5OO CHANNELS

T HE SAMOAN WARLORDS were in a kick line. Comcast, Continental, Cox, Newhouse Broadcasting, United Artists Entertainment, Viacom Cable and two companies that became Time Warner Cable all joined TCI in the Primestar coalition in 1990. The ownership was complex. The plan was simple.

"One of the biggest problems we've had over the years building cable is that somebody two miles from the end of the line goes to city council and says, 'They're doing a lousy job. I can't get service,'" said John Malone. "It was very helpful to us to have satellite to give to that guy." It was less helpful to rural cable operators, shut out of the Primestar partnership, who faced the service as a potential competitor. Malone had another plan for them.

Primestar was the first viable direct-to-home satellite TV service that didn't require a dish as big as a monster truck tire. The Primestar dish was three feet in diameter. It required a professional installation, which typically consisted of pouring 100 pounds of cement into a hole in the ground, anchoring a pole, attaching the dish, then connecting two coaxial lines from the dish through the house, to a set-top box and the TV. The equipment was built into a monthly fee ranging up to $35 per month for 10 channels, which consisted of seven superstations and three pay-per-view options. Primestar's corporate headquarters was in Bala Cynwyd, Pennsylvania, but each of the partners sold the service in their own territories. They avoided cable areas and headed out to the sticks and trailer park country. Few subscribers signed up for the mousetrap at first.

"The only reason the other cable guys came in was to keep an eye on Malone," said John Cusick, a former GE Americom executive who was Primestar's president from 1990 to 1995. "No one understood what Malone was doing."

The other members of the cable gang feared TCI might launch its own rat zapper and raid their territories. Primestar kept everyone with their hands above the table, while keeping the GE Americom bird away from Rupert Murdoch, Charlie Ergen, GM Hughes or any other prospective competitor. "Malone said Primestar was 'a way of inoculating the cable industry against DBS,'" Cusick recalled. "If he could sop up the 10 million rural homes, he could deny Hughes a lot of business."

Meanwhile, digital compression upped the ante. A Seattle company named SkyPix planned to launch 80 channels of digital television to homes by satellite, with 200 different pay-per-view movies each day. The backers of SkyPix were Fred and Richard Greenberg, two Harvard-educated brothers who sold limited partnerships in real estate, medical equipment and movie production in the 1980s. They were also flypaper for lawsuits, a curse that bedeviled SkyPix and helped lead to its demise. Hughes, Comsat Video Enterprises, the Home Shopping Network and Microsoft co-founder Paul Allen were all cited as investors, or potential investors, in SkyPix at various times.

At the National Cable Show in New Orleans in March of 1991, crowds gathered around a SkyPix demonstration of its video compression technology. Attending the convention, Malone said that the prospect of cable systems offering 200 to 500 channels by the mid-1990s was "highly realistic." His comment passed without much notice. But within two years, that vague possibility hardened into a gospel preached by Malone himself. The date on which 500 channels pierced cultural consciousness was December 2, 1992, at the Western Cable Show in Anaheim, California.

The show typically takes place inside a cavernous convention hall decorated to look and sound like a circus. Playboy bunnies, walking come-ons for the magazine's cable channel, sign autographs cheek-to-jowl with vendors hawking broadband coaxial rotary joints and double banana plugs with gold-plated solder terminals.

During the 1992 convention, Malone announced a TCI order for one million digital cable boxes, to be deployed in subscriber homes beginning in January of 1994. Primestar would go digital in 1993. With video compression, a 50-channel cable system could magically carry 500 channels.

"Television will never be the same," said Malone.

The New York Times carried the story on its front page the next day, with the headline "A Cable Vision (or Nightmare): 500 Channels." The generally positive article described the likelihood of TCI offering 50 college football games on a Saturday afternoon and allowing viewers "to order anything from pizza to jewelry by pressing a button on the television remote control."

The concept of 500 channels had arrived. It became shorthand for the technological possibility—and marketing myth—that Americans could watch whatever they wanted, whenever they wanted. Another phrase for it was "video on demand." But that was a pointy-headed term, just like "high definition television" and "digital compression." Five hundred channels was understandable. Lots and lots of channels. It even had a racy ring, like the Indianapolis 500. Freedom of choice. A jibe to Bruce Springsteen, who sang, "57 channels and nothing on." A couch-potato fantasy. The future in your living room.

Vaporware. That's the term in Silicon Valley for new software that may or may not materialize. Malone was doing the same thing, throwing the idea out there, testing the market, seeing what stuck. Never mind that cable companies, particularly TCI, lacked the engineering, programming and ability to deliver 500 channels, much less make a profit from it.

A media frenzy slowly built until 500 channels became manifest destiny on the electronic frontier. Industry observers flogged stories about the 500-channel future and the issues it raised: Where would the programming come from? How would viewers navigate all those channels? Do we really need 500? Isn't Malone brilliant?

Any questions about Malone's timing and motivation for making the announcement were shrouded in a fog perpetuated by the media coverage and Malone himself. His 1993 appearance on a PBS show, *TechnoPolitics*, was typical.

> INTERVIEWER TIM WHITE: Dr. Malone, there's been a lot of
> talk about the 500-channel cable system, but it's going to take a
> lot of time and effort for people to learn how to use 500 differ-
> ent channels. What's the really compelling argument for it, what
> they call the killer application?
>
> DR. MALONE: Well, Tim, I think that the 500-channel identifica-
> tion is kind of a misstatement. What we're really saying is we're
> putting in place facilities that would allow us to have up to, you
> know, 120 channels, and then multiplex those as much as ten
> times. We could be providing as many as 1,200 channels.

The killer application, Malone said, would be pay-per-view movies starting every 15 or 30 minutes.

And so it went.

"Five-hundred channels, movies on demand, interactive TV, home shopping malls—Malone is the one who will drive them into the market," wrote Kevin Maney in *USA Today* in 1993. "Malone's high-tech vision is not whimsical. It is firmly rooted in reality. Around TCI's glass and concrete tower outside Denver, employees call the boss Dr. Malone in deference to his long list of technology degrees."

Malone's intellect, academic background, market power and glibness were all employed to trumpet a 500-channel bluff amplified by a receptive media. Why? There are as many theories as promised channels.

"It was new and exciting," Maney said in 1999. "Malone knew which buttons to push, offering something people would want. He was both too far ahead of his time and operationally couldn't deliver. But you couldn't know that without crawling around his cable systems. It was a great sales job."

Malone's announcement came two months after Congress passed the 1992 Cable Act over the veto of President Bush. Since TCI's actions and Malone's antagonism toward regulation helped goad the crackdown, perhaps he felt the need to perform a cable industry facelift in public.

Or, he got carried away. "Malone is famous among cable engineers as being the eternal golden technology boy of the press despite the fact of never delivering," said Jim Chiddix, chief technical officer for Time Warner Cable. "It's purely for positioning."

Or, he got the timing wrong. After a series of delays from General Instrument, TCI did indeed install one million digital cable boxes in subscriber homes. It happened in 1998, four years after the promised delivery. The boxes provided 35 extra channels, plus 10 additional pay-per-view options, 10 audio music channels and an electronic programming guide. Not bad, but not 500.

Asked about the legacy of 500 channels in 1996, Malone said he brought the news of digital compression and that 500 was "just a number picked out of the air." That is a curiously vague statement coming from a man schooled in higher mathematics, credited with a four-dimensional mind and able to recall the details of hundreds of transactions.

In the winter of 1992, there was one critical reason for Malone to tie the fact of digital compression to the fantasy of 500 or 1,200 cable channels. It was this: The rat zapper would actually be able to deliver nearly 200 channels, four times the number of the average cable system. Digital compression could be deployed more quickly by satellite than through 11,000 cable systems. It's the difference between placing a searchlight in the sky versus changing all the bulbs in a string of Christmas lights across the country.

DirecTV and EchoStar were many months from launching DBS in late 1992. But Deathstar was coming and Malone knew it. Digital compression worked and the 1992 Cable Act mandated program access for cable competitors. Five hundred channels was a nice fat, round number to hang out in public, perhaps giving pause to investors who might think of financing DBS. In particular, 500 channels might warn mighty General Motors against launching service through its Hughes Electronics division.

That warning had already been delivered in person by J. C. Sparkman, the cable operations chief who made the cash flow numbers Malone needed to leverage TCI's growth. Well before the 500-channel buzz, Sparkman approached General Motors chairman Robert Stempel and reminded him that TCI and other cable companies were big buyers of GM truck fleets. "Why the hell do you want to compete with the cable industry?" Sparkman told Stempel. "We're some of your biggest customers."

Sparkman later said that his pressure resulted in TCI getting bigger discounts on their truck purchases from GM.

Stempel was ousted as GM chairman on November 2, 1992, as the automaker faced crippling business problems. A month later, as Malone's 500-channel announcement hit the news, GM announced it was closing nine plants and cutting 18,000 jobs, part of a 1991 plan to eliminate 74,000 jobs over five years. GM had reported losses totaling $5.4 billion in the previous seven quarters. Not an auspicious time to launch an untested satellite TV service against entrenched cable companies.

Malone had had opportunities to team up with GM Hughes on satellite TV dating back to the late 1980s, but shied away, knowing that the rat zapper could cannibalize cable. "I'll never forget it," said Cusick, Primestar's president, recalling a conversation with Malone about the GM Hughes satellite TV entry. "He said, 'I'm not convinced they're going to get their funding. And I'd have to be brain damaged to give it to them.'"

Malone's 500-channel announcement was the public face on a private strategy to blunt the rat zapper.

"John still believed as late as 1992 that there was a chance that DBS wouldn't happen," said Cusick. "With this half-assed Primestar, announcing 500 channels, leaning on General Motors and making it difficult for competitors to get programming, maybe he could discourage DBS from happening."

He may have slowed it down, but he couldn't stop it.

After the Sky Cable deal with Murdoch, NBC and Cablevision fell apart, the pursuit of the rat zapper at GM Hughes fell to Eddy Hartenstein, a company executive and former mission planner on NASA's *Viking* and *Voyager* space programs in the 1970s. Sporting prematurely snow-white hair and beard, Hartenstein also possessed an engineer's stiff bearing and diction. He was prone to discuss "augmenting install capacity" and "upside potential for revenue generation," which meant selling dishes and making money. But Hartenstein also had a wry sense of humor. He once called Primestar "Cable Helper"—like Hamburger Helper, a meal that would pale in comparison to the prime rib of digital DBS. Hartenstein and his team, including an engineer named Bill Butterworth and a programming executive named Jim Ramo, sought out partners for the service to ease the GM board's apprehension about footing the $1 billion start-up cost.

They approached the Hollywood studios and came away empty-handed. More likely prospects were the Hubbards of Minnesota, a family of firsts. Stanley E. Hubbard founded one of the first commercial radio stations in 1923, WAMD—"Where All Minneapolis Dances." He also started the nation's first commercial airline, which flopped. Hubbard owned the first independent NBC-affiliate TV station in 1948 and pioneered the first 10 P.M. local news broadcast. By 1960, Hubbard and his son, Stanley S., had bought the first color TV camera. In 1981, Hubbard Broadcasting formed United States Satellite Broadcasting (USSB), and was among the first applicants for a DBS license.

Stanley S. Hubbard was an early, vocal advocate of DBS. He drew hoots when he showed up at cable conventions with an 18-inch dish under his arm. He implored U.S. broadcasters to take it seriously and pointed to other countries. NHK in Japan sold more than 500,000 dishes for $750 apiece in 1987, offering just two TV channels. By 1991, millions of Europeans watched satellite TV, the largest share in Germany.

"The message that I'm giving to my friends in broadcasting is that if we don't do it, somebody else is going to do it," he said in 1987. "And if that happens, then the game is over and we're out."

While prescient, his view was somewhat narrowed by the broadcast business. He declined to buy into cable companies early on, because, he said, "We don't want to be riding on the back of other people's property," referring to cable's retransmission of broadcast signals. When the Hubbards won a DBS license, they requested just 16 channels—half of which they later gave back. The Hubbards were content with eight channels because they viewed DBS as a niche service for pay-per-view movies, culture and news, not a head-on competitor to either broadcast or cable. What good was having 32, or 300, satellite channels if they would never be filled with programming and never find an audience?

"My father saw DBS as a national television license, more like the advertising-supported broadcast business and not a subscription-supported business," said Stanley E. Hubbard, the president of USSB and son of Stanley S. "To do three or eight channels would be a major undertaking." The Hubbards missed out on a chance to snag an entire DBS slot, but they were savvy enough to ride others' coattails. When GM Hughes looked like it might launch service, USSB asked the FCC to transfer five of its channels to their location. "We wanted to be where the antennas are pointing. That's what was going to drive the business."

When GM Hughes came calling, the Hubbards agreed to kick in $100 million and operate their five channels in concert with the 27 held by Hughes at the 101 location, making up a full slot. USSB's job would be to sell premium channels like HBO and Showtime. The National Rural Telecommunications Cooperative also became a partner, putting up $125 million for franchise rights to sell the satellite TV service in rural America.

The service was called DirecTV. RCA-Thomson agreed to manufacture the satellite dishes and digital decoder boxes. Sony developed state-of-the-art videotape machines for DirecTV's $100 million digital uplink facility in Castle Rock, Colorado.

On December 17, 1993, a Hughes-built satellite blasted off atop an Ariane rocket from Korou, French Guiana. The satellite, with a 72-foot wingspan when deployed, achieved an orbit at 101 West longitude, traveling along a line from Bottineau, North Dakota, to Eagle Pass, Texas.

DirecTV officials celebrated in Los Angeles, while several hundred guests of the Hubbards partied at the Minneapolis Convention Center. Before the launch, a Comcast cable executive far away from the festivities mused about the possibility of firing a Stinger missile at the rocket and satellite as they rose in the sky.

IN JUNE OF 1993, the U.S. Justice Department ended its antitrust probe into Primestar, concluding that the cable partners had "engaged in a continuing agreement, combination and conspiracy to restrain competition" in the subscription television business. This alleged conspiracy came in two flavors. The Primestar partners had agreed to a "most favored nation" clause by which no other satellite TV service or cable competitor would ever get a better price on programming than did Primestar. Then, to enforce this price fixing, the partners would disclose to each other the price of programming sold to outsiders. Independent programmers that did not play ball might get a bad cable channel position, or no channel position at all. The Justice Department did not take its case against Primestar to trial. Nor did it seek to force the cable gang to divest Primestar for the sake of competition.

"A lot of theories were kicked around," said James Rill, then the Justice Department's antitrust chief. "I'm confident we did discuss the possibility of not letting them into the satellite TV industry. One of the reasons we were reluctant was that DBS was on the horizon."

The Justice Department filed a consent decree with Primestar's cable owners that basically echoed rules in the 1992 Cable Act that guaranteed competitors access to cable programming. Attorneys general in 45 states settled similar probes with Primestar in a separate consent decree. Just as with the 1992 Cable Act, the government was unwilling to take a decisive stand and take it to court if necessary. It didn't take an antitrust scholar to see that the cable gang's intertwined ownership of programming, cable delivery and Primestar was a potential chokehold on a truly competitive market. It would take another five years, and several rounds of cutthroat, before that issue was resolved—by the same Justice Department, with a different face at the top of the antitrust division.

HAVING VOWED 500 channels, Malone set his sights on engineering it. Or, failing that, engineering the sale of TCI. In early 1993, he announced that TCI would spend $1.9 billion to drive fiber-optic cable through 100 cities to deliver new services. TCI also spent $100 million building the National Digital Television Center (NDTC), south of Denver in Littleton. NDTC consisted of a capacious building fronted by satellite dishes like huge clams on the half-shell. NDTC was a clever addition to Malone's hedge against the rat zapper.

The purpose of the dish farm was to deliver digitally compressed video to large cable systems across the country and to Primestar customers. NDTC would also deliver digital channels and pay-per-view options by satellite to small rural cable systems. These new channels could be sent down the cable lines to subscribers without having to rebuild the cable system itself.

This could inoculate smaller cable systems against DBS, too, though some of the companies were uncomfortable ceding control of back-office operations and customer fulfillment to TCI. Primestar's Cusick called NDTC "Dr. No's basement," a reference to the James Bond nemesis who sabotages rocket launches from a Jamaica hideout. "John's plan was for everybody to get 200 channels of digital television all coming from Dr. No's basement," said Cusick. "He would be the guy with his finger on the switch."

What the Comcast executive joked about doing to DirecTV's satellite launch—that is, blasting the rat zapper—Malone was going to try to do subtly and without a midair explosion.

The chess machine lived to strategize, seeking ways to control the cable and satellite TV industries, just as Bill Gates sought to control the future of personal computers with the Windows operating system. Malone's endgame, however, was very different. He did not enjoy the day-to-day grind of running a cable company. He also saw that only the biggest media and technology conglomerates would survive in the twenty-first century. All the selling and building—the cable system acquisitions, the 500-channel announcement, the $1.9 billion rebuild, Dr. No's basement—were investments in the future, both offensive and defensive plays. They were also a neon sign in the new digital economy, blinking "Let's Make a Deal."

Ray Smith, the chairman of Bell Atlantic, was sold.

A $33 billion merger between TCI and the Philadelphia-based Baby Bell was announced on October 13, 1993. Once again, the superlatives flowed in the media. The merger was unprecedented, soul-shaking, evolutionary, revolutionary, historic, a paradigm shift and concrete confirmation of the information superhighway rolling across the land. It came at a time when phone companies sought to buy into cable companies to create communications titans, though most of the deals faded away. The Bell Atlantic/TCI merger was by far the most celebrated. Smith would run the merged company, while Malone devised strategy and remained the chief of Liberty Media, TCI's increasingly lucrative portfolio of programming assets.

The merger faced intense antitrust scrutiny. Adding to the pressure was a grenade thrown at Malone from within the cable-and-programming club by Sumner Redstone, the litigious chairman and chief of Viacom. Redstone had been a standout graduate of Boston Latin School and Harvard Law. He built his father's drive-in movie business into a 750-screen movie theater company and is credited with coining the term "multiplex." He also survived a 1979 hotel fire in Boston with a badly burned right arm. Redstone later led the 1987 leveraged buyout of Viacom, a company he quickly took public, building a programming powerhouse to rival Time

Warner and Disney. In addition to serving 2.5 million cable subscribers, Viacom owned MTV, VH1 and Nickelodeon. The company had dropped out of the Primestar alliance. It later bought
Blockbuster, the video store chain.

In November of 1993, Viacom filed a 91-page lawsuit against TCI
in which Redstone called Malone everything but Satan in Silk Paja-
mas. The lawsuit's opening paragraph had the crackle of a 1920s
newsreel: "In the American cable industry, one man has, over the
last several years, seized monopoly power. Using bully-boy tactics
and strong-arming of competitors, suppliers and customers, that
man has inflicted antitrust injury on plaintiffs Viacom and virtually
every American consumer of cable service and technologies.

"That man is John C. Malone."

The 19-count lawsuit asserted violations of the Sherman and
Clayton Antitrust Acts, with tortious interference and breach of
contract thrown in for good measure. Among the allegations was a
"conspiracy" between TCI and General Instrument to monopolize
digital compression technology. That triggered a Justice Department investigation into the relationship between GI and the cable
industry, a probe later dropped.

Some of Viacom's charges were *Godfather* hyperbole, like the
claim that TCI's Liberty Media made an "offer that Disney could
not refuse" of $1.5 billion for movie rights. Malone had not, after
all, put the severed head of Mickey Mouse into Michael Eisner's bed
as he slept. Rather, Liberty's Encore movie channel simply outbid
Viacom's Showtime for the Disney movies. The suit went on to cite
unnamed TCI officials as vowing a "crucifixion" of Viacom's other
pay service, The Movie Channel, by dropping it in favor of Encore
on TCI cable systems. Viacom further alleged that TCI used its control of the cable pipe to demand stakes in programmers such as the
Learning Channel and Court TV.

What was Redstone's motivation? At the time, QVC, a home
shopping channel led by Barry Diller and owned primarily by
Comcast and TCI, was in a nasty bidding battle with Viacom for

Paramount Communications. Redstone's lawsuit was an attempt to gain leverage by exposing TCI as a rapacious bully. But with the Bell Atlantic deal, TCI backed off supporting QVC in its bid for Paramount, which Viacom eventually won.

Malone was said to be stung by the personal nature of Redstone's lawsuit. It fueled his desire to recede from the public eye, which was one of his main personal motivations for the Bell Atlantic merger. Though a Bell Atlantic/TCI union would have a high profile, Malone planned to stay in the background, focusing on strategy and Liberty Media. He would regain the privacy he and his family had lost when he began building a cable empire.

"In a world where you have to hire private detectives to protect your kids and you have to worry about going out of town at night because people know where you live, where your family is a target for all kinds of things, sure, you want to be low profile," he said in 1993. "You don't want to be in the newspaper every day. You don't want to have people printing a map to your house, which has happened to me. In this world, frankly, the less visible you are, the better."

Malone's financial engineering of the cable pipe and programming helped advance the entertainment culture, but he himself was an unwilling celebrity at a time when captains of the Information Age were covered like rock stars. Malone was famously prickly about interviews, declining most requests and even stiffing some he had agreed to. The view from inside TCI was that Malone could be painfully shy, which was true; that he generally distrusted the news media, also true; and that he was sometimes naive in handling the press, which was laughable.

When he had something to sell, be it 500 channels or the Bell Atlantic deal, Malone could put the latest Hollywood starlet to shame. Flashing cleavage of a different sort, Malone could stand before a group of reporters and riff for an hour off the top of his head, barely listening to questions, just downloading from the supercomputer atop his shoulders as the scribes took in the view. Even

Charlie Ergen, who stayed up nights worrying about Malone's next move, struggled to keep up. "He talks at 60 miles an hour. My mind works at 40 miles an hour, so it's a little hard for me to follow."

The same control dynamics and skills Malone brought to business negotiations were evident in his press relations. Malone once declined a one-on-one interview, telling the reporter, "You need to write better stories."

The Bell Atlantic/TCI merger died stillborn, four months after it was announced. Malone and Smith could not make it work. They could not find a way to get the Baby Bell's dividend-seeking investors to climb up TCI's debt-leveraged pyramid to see the view. Nor could they strike a mutually agreeable merger price, particularly after the FCC had cut cable rates 10 percent in late 1993 and an additional 7 percent in early 1994.

Malone's dream of stepping into the background would have to wait for another day. Less than 24 hours after he and Smith pulled the plug on the deal of the decade, Malone was back on the stump, looking for the next move to make.

"I've got my uniform back on and I'm ready to play," he said at a TCI press conference. "Put me back in coach." Malone described his wife as "disappointed" by the decision not to pursue the merger. "But I said I can't retire that way now."

———

THE FAILURE OF the Bell Atlantic/TCI merger did not dampen the enthusiasm of local phone companies to offer cable-like video services direct to homes. Indeed, some saw it as a necessity to become a big wheel on the much-hyped information superhighway. A *Cablevision* magazine cover story in 1995 described the Baby Bell strategy as "The Plot to Cripple Cable."

The Bells were blocked by federal law from owning monopoly cable companies in their own territories. But they could try to run video down their own copper phone lines, a technology called video dialtone. It didn't work very well. Or, they could build a com-

petitive cable network, which Ameritech did in the Chicago suburbs. That was expensive. There was also wireless cable technology,
which sent TV signals from radio towers to subscriber homes. Hollywood deal maker Michael Ovitz put together Disney and three
Baby Bells—Nynex, Bell Atlantic and Pacific Telesis—in a wireless
cable company called Tele-TV, headed by Howard Stringer, a former CBS chief. It didn't last long.

US West, the Colorado-based Baby Bell, was the most aggressive
in buying cable assets outside of its phone service territory. It spent
$2.5 billion for a stake in Time Warner Entertainment in 1993 and
another $11 billion to buy Boston-based Continental Cablevision in
1995. It ended up spinning those assets off into a separate company.

The Bells were eager to buy into cable, or co-opt it, but they did
not end up competing with it very much in the 1990s. The threat to
cable came from above.

—

IN MARCH OF 1994, Primestar went digital, expanding to 75 channels. This came after the Justice Department consent decree and
much internal debate among the cable gang about how to proceed.
"Time Warner was terrible," said Cusick. "They didn't want to go
digital. They didn't want to go high power. They were very
obstructionist."

Even with its mixed motivations, Primestar began to take off,
gaining 300,000 subscribers by the end of 1994. They launched a
$100 million national advertising campaign in 1995. "We had one
million calls in 30 days," said Cusick. "But 60 percent of the calls
came from cable areas. They went straight in the wastebasket. We
were too successful. People were too damned conflicted. They
wanted to be in the business, but they didn't want to be in the
business."

Cusick said his error was throwing in his lot "with a bunch of
guys who didn't want to see it work. I spent eight years of my life
doing that. At some point, you're just burnt out." Before Cusick

resigned from Primestar in June of 1995, he discussed its fate with Malone, whose company was the most aggressive of the Primestar partners in selling the service.

"This has to be a pretty good business," said Malone. "The thing's been shot, stabbed and poisoned and it keeps getting up."

The pressure on Primestar and the cable gang came from DirecTV. In June of 1994, DirecTV launched a second satellite to the 101 slot, and later, a third. Several satellites could be launched to a particular DBS slot, but that did not expand the number of transponders, or channels, licensed for use by the federal government. Channel expansion did come with digital compression, which squeezed 10 or more channels onto each transponder.

All the technology and satellites meant nothing, however, if subscribers weren't convinced to plunk down between $699 and $899 for DirecTV equipment, plus installation and monthly programming fees.

"The business plan cannot be a technology in search of a market—it had to be competitive," said Eddy Hartenstein, the company's president. "I think a lot of people in the cable industry dismissed us. That's fine. We really didn't feel a need to climb on a table and crow about how we were going to compete with cable."

By the time of DirecTV's rollout, having played his best defense against the rat zapper, Malone raised a toast. "We have an ownership interest in 44 of their channels," he said. "So if they do well, we do well."

Murdoch, burned twice on DBS in the United States, was less sanguine.

"I don't want to discourage the entrepreneurs who are pursuing that course," he said, just months before DirecTV dishes hit the market. "But I would say that in the creative industry, or Hollywood if you like, the overwhelming majority of people are skeptical. Because the cable is in the ground. If people want extra choice, they are already capable of getting it."

Murdoch missed a key factor: the pent-up demand for an alternative to cable. DirecTV was a service whose time had come, taking advantage of high-powered satellites, mini-dish technology, digital compression and program access rules. Jackson, Mississippi, was one of five initial markets for DirecTV in the summer of 1994. "Every time we get units in, they're sold out the same day," said Edward Maloney at Cowboy Maloney's in Jackson. "I would have no qualms about taking a big shipment of 5,000 units right now because the demand is that big."

DirecTV sold one million systems in just over a year, far more than the number of VCRs, CD players and TVs sold in the same time frame when they were introduced. Within four years, DirecTV would claim four million subscribers, one of every 25 U.S. homes, which placed it in a league with the nation's top cable companies.

DirecTV provided what cable and Primestar could not: 175 channels of digitally sharp pictures and sound. Well-funded, well-timed and market savvy, DirecTV was the airborne assault Malone and the cable gang had held at bay for years. DirecTV dwarfed the number of channels of most cable systems, offering pro sports packages and specialized networks ranging from the Golf Channel to TV Asia. Three feeds of HBO were available, which most cable viewers didn't even know existed. DirecTV set aside 55 channels for pay-per-view movies, with half-hour start times, at $3.99 a pop. This home video store approach was what Malone had called the killer application.

DirecTV also beat out cable in customer service. Hartenstein selected the company's seven-letter name, preceded by an 800 prefix, so it could also serve as a toll-free customer service phone number. DirecTV manned its phones 24 hours a day. "It was a pretty sorry state of affairs that Congress had to dictate to the cable industry specifications on how to answer their phones" in the 1992 Cable Act, said Hartenstein.

Hughes poured more than $80 million into marketing and advertising DirecTV its first year, a unified national campaign that the fragmented cable industry had never themselves done, but which Primestar emulated. The press had a story of satellite innovation versus cable monopoly and they jumped all over it. Stories on

DirecTV appeared in publications ranging from *Newsweek* to *Electronics Now*, from *Esquire* to *Motor Boat & Sailing*, which hyped the DBS service as a new toy for ships at sea. DirecTV won over rural subscribers in uncabled areas and home electronics nuts everywhere, markets described as "low-hanging fruit" by Hartenstein. Yet, it also lured city dwellers who could place the pizza-sized dish on high-rise terraces. More than one-quarter of DirecTV's subscribers were in cabled areas, many of them willing to spend $60 or more a month for a new breed of television. By luring the highest-revenue customers, DirecTV put the cable industry on notice that its best wasn't good enough anymore.

John Malone controlled the nation's largest cable company, TCI. He had a flanker satellite TV strategy with Primestar. But to counter the momentum of DirecTV and the hard-charging EchoStar, Malone needed a full-fledged rat zapper himself. What he needed, but did not have, was the last full DBS location in the United States: the 110-degree slot.

Ergen wanted it, too. So, it turned out, did Murdoch.

The man who had it was Dan Garner.

EIGHT

FIGHTING

FOR AIR

D AN GARNER IS A good old boy. He stands six-foot-three and weighs 230 pounds, with a two-pack-a-day cigarette habit, alternating between Marlboro reds and Merits. He claims to be one of the youngest Eagle Scouts in the history of Arkansas and can recite the 12-point Scout Law off the tip of his tongue: "Trustworthy, loyal, helpful, friendly, courteous, kind, obedient, cheerful, thrifty, brave, clean and reverent."

A bachelor in his mid-50s, Garner spends his days in a Washington, D.C., apartment. He plays the stock market, watches football games on TV and entertains lady friends "as often as possible." His mission, however, is chasing the ghost of 110 through the courts.

The man who would be king of satellite TV is down-home. "I'll go home before I'll get a horsewhipping," Garner says about a bad business deal. Or, "I've had a couple of pops," when conversing baldly after a few cocktails. Or, "Gee whiz shucksaroo," to connote mock amazement.

But when talk turns to 110, Garner's tone becomes deliberate, profane and pained. He once held the DBS slot, and the hundreds of millions of dollars it represented, in his hands. It slipped away in a politically charged game of cutthroat.

"I was like the monkey screwing the skunk," said Garner. "I didn't get all that I wanted, but I got all I could stand. I ... got ... ALL ... I ... could ... stand."

Garner was among the first in line for a DBS license.

The first group of 14 applicants included Comsat, Dominion's Robert Johnson and the Hubbard family. A company called Home Broadcasting Television Partners of Milwaukee was disqualified because "Home" turned out to be a prison in Terre Haute, Indiana, where applicant Richard Wagner was serving time for loan fraud.

The second round of applicants, in 1984, included Garner.

He didn't know much about DBS, but he was a quick study and a sharp-eyed opportunist. He'd been president of his freshman, sophomore and senior classes at the University of Arkansas at Little Rock. He won the last election with the slogan "If you vote for me this time, you'll never have to vote for me again."

He put himself through college promoting concerts and dances after Arkansas Razorbacks football games. After college, he booked groups like Chicago and Grand Funk Railroad into the Barton Coliseum, drawing crowds of 12,000 to the Little Rock venue.

In 1971, Garner started the city's first all-rock, FM radio station, KLAZ, and dabbled in real estate. He sold the station in 1976 and began prospecting in communications. Working out of his Little Rock office, he applied for broadcast TV, radio and cellular licenses at the FCC in Washington. He lost out on TV and cellular. But he won local radio licenses, in part, by finding minority and women partners, then selling the licenses at a profit. "A female minority is going to have the inside track over a bunch of white boys," said Garner, referring to FCC minority preference rules at that time. The other license Garner sought, which didn't require minority participation because there were so few takers, was DBS.

Garner knew about the business mainly from reading the trade press. He saw it as a potential competitor to cable, but moreso as beachfront property in outer space.

"Orbital real estate," Garner called it.

He connected with Don Dement, who headed NASA's advanced communications project on DBS. And with Gordon Apple, a TRW engineer. Dement and Apple understood early on that digital compression would make DBS a powerful business opportunity. They tutored Garner on the technology. He renamed his company from Garner Communications to Advanced Communications Corp. and ordered two RCA satellites in 1986.

"Dan had a license, which was valuable," said Apple. "But he had no money for development. Very seldom did anyone get paid. And Dan was not easy to work with."

The DBS license was free, but required engineering studies, lawyer's fees and seed money for the satellites, which ran into hundreds of thousands of dollars, a tough nut for Garner, who described himself as "a broke-dick son of a bitch."

He had some connected friends in Arkansas, including Jackson T. Stephens Jr., a scion of the Stephens financial conglomerate in Little Rock. Stephens once played keyboards in a band named Rayburn that Garner hired as opening act for a Three Dog Night concert in 1971. Stephens extended a line of credit to Garner for his DBS start-up, which Garner said he used sparingly.

"I considered making a substantial investment in this," said Stephens in 1999. "For Dan's sake, I wish I had."

Garner also came to know Wilbur Daigh Mills, the legendary Democratic congressman from Arkansas who ran the House Ways and Means Committee. Mills was stripped of his power after two public peccadilloes with Argentine stripper Fanne Foxe in 1974. Mills won reelection that year and served out his term.

In 1988, as a private citizen, Mills founded the Foundation for Educational Advancement Today in Arkansas. Garner convinced Mills to become chairman of his company, while Garner committed one-eighth of the satellite slot's capacity to the foundation's work: Every American school and library was to get a free satellite dish. Garner also envisioned selling music direct-to-home by satellite, but the main moneymaker would be subscription satellite television.

There were big dreams that central Arkansas might become an offshoot of Silicon Valley. Garner lacked only a deep-pocketed partner. "I talked with Disney, Universal Studios, Paramount, ABC, CBS, NBC, the Tribune company and manufacturers," Garner recalled. "Anyone that could conceivably have an interest in putting some money into an engine to pull this long train of services."

Garner's Washington law firm, Hogan & Hartson, contacted Rupert Murdoch, to no avail. Garner got close to a deal with Bernard Schwartz, the chief of Loral, which built satellites. He struck out there, too. To keep Advanced going, Garner mortgaged two homes owned by his mother in west Little Rock. He lost both. The first foreclosure was overseen by an attorney and creditor named Jim Guy Tucker, who succeeded Mills in Congress and later became the state's governor.

After years of FCC policy changes, many of them driven by the wreckage of DBS start-ups from the likes of Comsat and Murdoch in the 1980s, Garner hit paydirt. In late 1991, Advanced was awarded 27 of 32 channels at the 110 slot. With digital compression, that could mean 270 or more channels. Garner owned the hot slot.

Yet, only one suitor realized it and pursued Garner with any zeal. Only one saw that Garner's DBS license was a license to print money.

Charlie Ergen wanted 110 badly, though he was no further along than Garner in launching DBS. Ergen was selling big-dish satellite equipment domestically and overseas, grossing $200 million in annual sales. But he had not applied for a DBS license until 1987 and was awarded just 11 of 32 channels at the 119 slot in 1992. If Advanced and EchoStar joined forces, they could marry the 110 and 119 slots and launch a mass-market rat zapper. Garner and Ergen needed each other. In some ways, they deserved each other. They were slick talkers who came from the same region of the country and didn't trust one another.

"Hardworking, bright, dedicated and slippery when wet," Garner later described Ergen.

"He was a genius who thought outside the box and knew how to beat the system," said Ergen of Garner. "But Dan's biggest enemy is himself."

The two men talked over many months beginning in 1989 and finally shook hands on a deal during a July 1992 meeting at Dulles International Airport outside Washington, D.C. They signed a letter of intent that month to form a fifty-fifty joint venture. Garner would be chairman and chief executive; Ergen, the president.

"Good old boys from Tennessee and Arkansas should join forces," said Ergen, making a reference to their respective home states and to those of the Clinton/Gore presidential ticket that year.

Ergen paid Garner $1 million as a show of good faith. The check bounced. Ergen sent another, which cleared. Garner was engaged to Ergen. But he was also playing the field. In mid-1992, he got a letter from John Cusick, Primestar's president. "It has been quite some time since we last spoke and it seemed to me that it may be useful for us to resume our conversation of last year," Cusick wrote.

Such approaches are commonplace. Satellite TV, like cable, is run by a small group of men who have crossed paths and swords many times. Discussions of partnerships, alliances and buyouts are constant to minimize the risk of the stomach-churning, billion-dollar investments they must engage in to stay in the game.

Ergen chased Garner. John Malone chased Garner. And Garner ended up chasing his own tail.

Between 1992 and 1994, Ergen and Garner debated the terms of their joint venture: the tax consequences, the need to get satellites built quickly and the use of the $1 million payment. Each man wanted voting control of their joint company and they could not come to terms. "He kept fucking around back and forth," said Garner. "I'm sure I did for a couple of years, too."

On December 14, 1993, Garner wrote a letter to Bert Roberts, the chairman of MCI, proposing a DBS partnership. No response. An MCI executive later said it was "one of the most bone-headed decisions we had ever made."

———

PRIMESTAR WAS AT war with itself. TCI/Tempo won preliminary approval from the FCC to launch DBS service to the 119 orbital slot in 1992. But the license was only for 11 of 32 transponders. And, the license was bound up in the Primestar partnership. The company's board of directors could not agree on whether it was worth the expense to launch a satellite to that slot. They also could not agree on whether Cusick should pursue an agreement with Garner to buy the more favorable, adjoining slot at 110.

Frustrated at the indecision, Malone sent TCI executive David Beddow, Primestar's former chief operating officer, in to negotiate with Garner.

Time was running out. The FCC had announced that the days of DBS pioneering were over. Garner needed to get a satellite launched or risk losing his license. His political juice in Washington had evaporated with the death of Wilbur Mills in 1992. After a decade of development, Garner faced a cold reality: either a rocky partnership with Ergen or selling out to Malone. He played both sides against the middle and got squeezed.

In August of 1994, Garner applied to the FCC for a four-year extension to construct, launch and operate DBS service at 110. His license was due to expire in December.

On his way to Europe for a business trip in September, Ergen visited Garner at his Washington, D.C., apartment. The residence was a symbol of their joint venture. Garner had moved from his one-bedroom apartment to the two-bedroom place so Ergen would have a place to stay when he came into town. "I switched and got a lease for another place, and we bought some furniture and tried to make it nice, because he brought his wife to town and stuff," said Garner.

When Ergen arrived at the apartment, Garner relayed some bad news: "I told him that I was heartbroken that we never could get a deal done. But it just had gone on and on and on, and we never could get there. So I sold to TCI."

Ergen grew angry. "Well, that means you're going to pay me the money back," he said, meaning the $1 million advance. Garner cited expenses, including $230,000 for legal fees, lobbying costs and the apartment lease. According to Ergen, Garner thanked him for helping Advanced stay out of bankruptcy and keeping the 110 satellite license thus far—a license that Garner was about to sell to Malone for more than $50 million.

"I took it as he was almost mocking me for being so stupid," Ergen said later. Garner recalled his own reaction as "I'm really sorry and maybe one day we can do something else together."

Ergen and Garner were like a weary couple whose overlong engagement comes to a bitter end before they reach the altar. Garner cooked a meal and they had something to eat. Then Ergen went to bed in the spare room. Both men got up around 5:30 A.M. Garner questioned Ergen about whether he would sue. Ergen said he might.

"Well, will you give me, will you call me before you sue me?" Garner asked. "There's no point in us getting in a lawsuit that's ridiculous."

Within weeks, Garner was served with papers in a six-count civil suit that alleged breach of contract. It demanded the return of the $1 million and sought damages in the billions. On the day the lawsuit was filed—September 29, 1994—TCI announced that its subsidiary, Tempo DBS, had signed a deal with Advanced and Garner for the coveted 110 satellite slot. "TCI is glad to have played a part in assisting Primestar in finding the right solution for growing its business," Malone said in a press statement.

He had sandbagged Ergen again.

As had Garner, who said, "It is appropriate for a visionary company like TCI and its subsidiary to acquire our company's DBS license." This was a far cry from Garner's statement six years earlier, when Advanced filed an objection to TCI getting into DBS at all, citing its "demonstrable track record of cutting off competition." Garner said later that the 1988 broadside, filed at the FCC, "was strictly a ploy for financing, to get someone else interested in what we were doing. I'd say, 'If TCI thinks DBS is so valuable, then why don't we make money on it ourselves?' Just words."

It was Malone, through Beddow, who ended up speaking the right words. Garner would get two million TCI shares, worth about

$47.5 million at the time, plus a onetime payment of $600,000 and a $30,000-per-month consulting contract for several years. Advanced would get to keep two satellite transponders at the 110 slot for its educational services. Garner, who had never met or spoken to Malone, was sitting pretty. And Malone had his high-powered rat zapper to compete against DirecTV and the soon-to-launch EchoStar. He had two Loral satellites waiting in the wings.

Garner had already applied to the FCC for a four-year extension to initiate service at 110. He then applied to transfer the slot to TCI's Tempo. All he needed was a rubber stamp.

IT TOOK NINE MONTHS for Ergen's lawsuit against Garner to go to trial in Denver. During a deposition, Garner's Arkansas attorney, Robert M. Cearley Jr., lured Ergen to restate his demand for the $1 million.

"Pay me the million bucks back, then [Garner] could go talk to anybody he wanted to," said Ergen. Whereupon Cearley handed over a cashier's check for $1 million made payable to Ergen and drawn on the Riggs Bank of Washington, D.C.

"I have to go to the bathroom," said Ergen.

"Hang on to that," piped up Ergen's attorney, a Texan named T. Wade Welch. "Make sure it doesn't get lost."

Ergen took the $1 million check, but pursued the lawsuit against Garner, seeking interest on the loan, plus damages. A six-member jury found for EchoStar on one count only, awarding $1, plus $235,000 in interest. Both sides declared victory.

By that time, the game for 110 had attracted some heavy action. Joining Garner, Ergen and Malone at the table were FCC chairman Reed Hundt, Rupert Murdoch and MCI's Bert Roberts. MCI failed to catch Garner's 1993 pitch for a DBS partnership. But Roberts did hook up with his old friend and MCI financier, Michael Milken. After his jail term, Milken started over in Los Angeles, teaching at UCLA's business school. One day in 1993, Milken asked Roberts to be a guest speaker. When the two men arrived at class, Rupert Murdoch was there to greet them. Roberts and Murdoch hit it off and pursued the lure of the Information Age: synergy.

The idea was that News Corp.'s programming content would flow through MCI's phone pipes, while together they planted their flag on the frontier of satellite, Internet and communications services.

Under founder Bill McGowan, MCI had scraped and clawed its way into competition with AT&T, first by reselling AT&T's long-distance phone service and then building its own network to compete with it. Just like the cable guys did with the broadcasters.

When McGowan died of a heart attack in 1992, his protégé Bert Roberts became MCI's chairman and chief executive. MCI became a big swinging pipe with enough hype to rival Malone's 500 channels. A full-page MCI ad in *The New York Times* on January 5, 1994, ended with this modest assertion: "The space-time continuum is being challenged. The notion of communications is changed forever. All the information in the universe will soon be accessible to everyone at every moment. And all because of a dream known as the Information Superhighway and a vision known as networkMCI."

In May of 1995, MCI invested $2 billion for 13.5 percent of Murdoch's News Corp. Much of their alliance, including an Internet service called Delphi, proved fruitless. Murdoch, however, had shown himself adept at deploying DBS in the United Kingdom and Asia, regions where cable lagged. Milken was keen on satellite technology, too. Yet, no full DBS slots were left in the United States. Malone had snagged the last one from Garner. Or had he?

Roberts became fixated on 110. And Murdoch, according to one former MCI executive, "needed a nice, fat, white-bread company" to front yet another DBS venture in the United States. MCI fit the bill.

Based in Washington, D.C., MCI is a battle-scarred veteran of the federal bureaucracy and a deep-pocketed campaign contributor to both sides of the aisle. "Government is a reality of life," McGowan once said. "Denying it is just letting your biases influence your business judgment."

On this issue, MCI's entrepreneurial corporate culture was dia-metrically opposed to that of TCI. And Bert Roberts was ready to storm Capitol Hill to seize the 110 slot from Malone. "The last thing this company was going to do was let the cable industry tie up satel-lite capacity where we would have a nice warm little arrangement where the [Baby] Bells would have their business and the cable com-panies would have their business and they wouldn't compete against each other," Roberts told *The Washington Post Magazine*.

Roberts's aggression came at an opportune moment.

By 1994, the FCC was the province of Reed Hundt, a former an-titrust attorney who owed his position primarily to Vice President Al Gore. Hundt worked on Gore's unsuccessful presidential pri-mary campaigns in 1988 and 1992, and was an economic adviser to the victorious Clinton/Gore campaign in 1992.

Hundt and Gore were classmates at St. Albans in Washington in the 1960s. Both attended Martin Luther King Jr.'s "I Have a Dream" speech in 1963 on the Washington Mall. Hundt's wife, a psycholo-gist, later described her husband as a "child of the sixties." Hundt graduated from Yale, taught for two years, then entered Yale Law School, where one of his classmates was Bill Clinton. Appointed as FCC chairman two decades later, Hundt joked, "I owe this job to lots of hard work and to fortunate seat assignments in high school and law school."

Critics saw his elevation to FCC chairman as a misfortune. Like his patrons at the White House, Hundt said all the right things about free speech, free markets and competition. But when it came down to cases, Hundt gloried in government's role as social engineer.

"He was like Attila the Hun, just bent on controlling everything," said FCC commissioner James Quello, nicknamed "The Boss" by his staff. Quello was born in 1914, served in the army during World War II and had been a broadcaster in Detroit. A Democrat, Quello was appointed to the FCC in 1974 and was acting chairman when Hundt came on board. No stranger to politics, Quello said later that Hundt's tenure was the most politicized he had seen in the agency in two decades.

Rachelle Chong, a Republican lawyer appointed to the five-member commission in 1994, said, "I felt very strongly that Gore had an agenda and that Reed was placed there to fulfill that agenda. I don't think he hid that."

The FCC is billed as an independent agency and nonpartisan, which is a fiction. Its budget is appropriated by Congress and its policy cues come from the White House. The agency's five commissioners are appointed in a 3–2 composition favoring the party of the president.

The agency began life as the Federal Radio Commission in 1927 to bring order to the conflicting claims over radio spectrum. In the process, it nationalized ownership of the airwaves and gave to the government powers that could have been left to the free market, balanced by the courts. The FCC often became a hindrance to the development of new communications technologies and competition, not a help. A telling anecdote involves the agency's 1954 decree blocking the sale of the Hush-A-Phone, a small metal cup that attached to the mouthpiece of a phone to provide privacy. The FCC declared it a "foreign attachment" competing with Ma Bell. As communications companies grew larger and the issues more complex, the FCC became less like the umpire of a tennis match and more like the judge of a mud wrestling free-for-all.

Into this ring stepped Hundt, who intended to make his mark. "Reed was a passionate guy not beholden to any industry interest," said Judy Harris, hired by Hundt in 1994 to head up the FCC's office of legislative affairs. "We had a motto, and had t-shirts printed up, that said: 'Read the law. Study the economics. Do the right thing.'"

One of Hundt's first orders of business was cable TV rate regulation. On Quello's watch, the FCC enacted 10 percent rate cuts in 1993, following the dictates of Congress in the 1992 Cable Act. Hundt led a second round of cuts, adding a 7 percent slice in February of 1994 that was credited by some with causing the demise of the Bell Atlantic/TCI merger. On the job just three months, Hundt celebrated with a speech at Harvard University, cited as the first annual Action for Children's Television Lecture on Media and Children. The speech, as prepared for delivery, included these remarks:

> Later last week I read in the newspapers that our cable rate decision allegedly broke up the Bell Atlantic–TCI merger, making me and my fellow commissioners the biggest trustbusters since Teddy Roosevelt.
>
> I appreciate the compliment, but I think it may have been undeserved.
>
> However, it is true that we at the FCC have a fair amount of responsibility for developments relating to the greatest story in the history of communications since the invention of the printing press: the National Information Highway. For the record, that term was coined in the late 1970s by a graduate of the Harvard Class of 1969: a first-term congressman named Albert Gore, Jr.
>
> Incidentally, I hope someone's counting my references to Cantabridgians: I'm going for a record here. After 30 years of trying to get into Harvard, I can't count on returning.

There was his erudition—the reference to Cambridge residents as Cantabridgians; the bow to his patron, Al Gore; and his conceit—with the "undeserved" reference to Theodore Roosevelt.

Time Warner's Gerald Levin called the agency's cable rate cuts "Soviet-style regulation."

Malone suggested a Stalinist solution. It came in a July 1994 *Wired* magazine interview, conducted shortly after the collapse of TCI's merger with Bell Atlantic.

Malone was caricatured on the magazine's cover as the "Infobahn Warrior," dressed in leather and carrying a sawed-off shotgun like Mel Gibson in *The Road Warrior*. The interview by David Kline began with Malone pointing to a stuffed gorilla dressed in a vest and tie in his office. He referred to it as Ray Smith, Bell Atlantic's chairman. Malone critiqued the Baby Bells and alluded to software deals with Bill Gates. The interview ended with Malone making a half-baked vow that TCI would quickly build out its sections of the info highway.

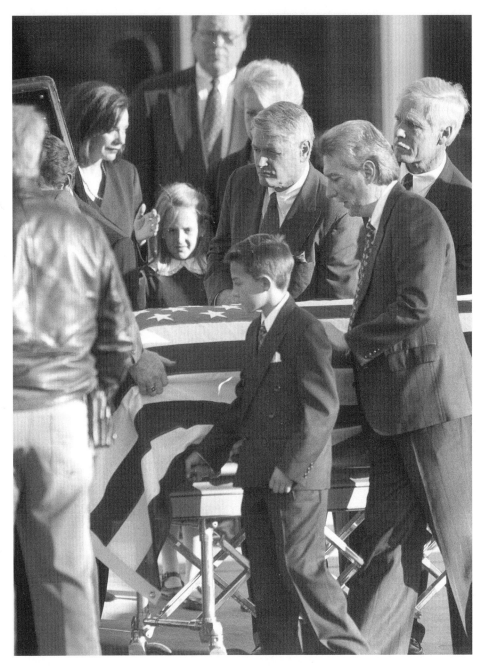

Clockwise from center: pallbearers John Malone, Ted Turner, Kim Magness and his son guide the casket bearing the body of TCI founder and chairman Bob Magness to a waiting hearse on November 20, 1996. Malone, TCI's mastermind, could not foresee the turmoil about to engulf the world's most powerful cable TV company. (David Zalubowski, Special to *The Denver Post*)

Charlie Ergen described himself as a "country boy from Tennessee" just selling satellite dishes. But his partnership with Rupert Murdoch in 1997, nicknamed Deathstar, would rock the industry, provoke the cable gang and nearly lead to the downfall of Ergen's company, EchoStar. (Dave Buresh, *The Denver Post*)

Charming and conniving, Rupert Murdoch intended to supply the cable gang with Fox programming, while stealing their subscribers by satellite. "I'll bring them to their knees," he said. (AP/Wide World Photos)

Michael Milken, the dethroned junk bond king, and adviser to moguls, told Murdoch and Ergen in 1996 that their partnership would "change the world." (AP/Wide World Photos)

TCI deal maker John Sie saw the cable industry as a "total neophyte" amid the power politics in Washington, D.C. (Crissy Pascual, *The Denver Post*)

"My God, what an invention is this!" said Bill Daniels when he first saw TV in 1952. He became the cable industry's most influential broker. He is pictured *(above right)* in 1997 with Daniel Ritchie, chancellor of the University of Denver, a former media company chief and an executor of the troubled Magness estate. (Kent Meireis, *The Denver Post*)

Malone confidant Peter Barton ruled Liberty Media, TCI's programming company, until 1997. (Andy Cross, Special to *The Denver Post*)

A sailor in his spare time, John Malone (pictured here in 1982) built TCI into a cable-and-programming powerhouse through economies of scale, debt leverage and tactics, both subtle and strong-arm. (Ed Maker, *The Denver Post*)

As a rising U.S. senator from Tennessee, Al Gore *(left)* was a self-appointed prosecutor of what he called the "Cable Cosa Nostra," with Malone its unscrupulous head. He helped pass the 1992 Cable Act, which sharply cut rates, before becoming vice president under Bill Clinton, with whom he is pictured in 1999. (AP/Wide World Photos)

A good old boy from Arkansas, Dan Garner had a billion-dollar satellite TV license in his hands, only to have it snatched back and auctioned off by the Federal Communications Commission in 1996. (Barbara Ries)

Erudite, passionate and pompous, Gore acolyte Reed Hundt led the FCC from 1993 to 1997. The agency's cable rate cuts led Malone to suggest, "Shoot Hundt!" (AP/Wide World Photos)

Arthur C. Clarke, co-creator of *2001: A Space Odyssey*, published a theory in 1945 that envisioned satellites orbiting Earth, transmitting TV and data signals. His idea, which initiated a global industry, earned him $40. (AP/Wide World Photos)

Microsoft chief Bill Gates clashed with TCI's John Malone as software and cable technologies converged. "They'd gotten themselves under the gun in a lot of ways," Gates said of the cable industry. (Karl Gehring, *The Denver Post*)

On the ropes after Deathstar's sudden death in mid-1997, Charlie Ergen's EchoStar took on more debt to launch a third satellite, while suing Murdoch's News Corp. for $5 billion. (Dave Buresh, *The Denver Post*)

As much a politician as a cable TV executive, Leo J. Hindery Jr. *(center)* took over as president of TCI in early 1997 and whipped John Malone's troubled company into shape. With Hindery at the Western Cable Show in 1997 were media moguls Barry Diller *(left)*, and Ted Turner *(right)*. (AP/Wide World Photos)

Bob Magness and his second wife, Sharon, connected through a love of Arabian horses. But Magness's death in 1996 triggered a battle for his $1 billion estate among his widow, his two sons by his first marriage, and his longtime business partner, Malone. (David Zalubowski, Special to *The Denver Post*)

Malone's gaze hid a mind that worked like a chess machine. In the first half of 1997, he had many pieces to move: restructuring TCI, undermining Deathstar and controlling the Magness estate's TCI stock. (Shaun Stanley, *The Denver Post*)

The Family Channel, founded by televangelist Pat Robertson, was used by co-owner John Malone as a bargaining chip to lure Murdoch out of Deathstar. (AP/Wide World Photos)

Joel Klein, chief of the U.S. Justice Department's antitrust division, approved Baby Bell mergers, while targeting the cable gang's Primestar alliance with Rupert Murdoch as anti-competitive. (AP/Wide World Photos)

Carl Vogel joined EchoStar in 1994, helping Charlie Ergen launch the company's cut-rate satellite TV service. Vogel later became chief of the cable gang's Primestar, which was doomed to failure. (Helen H. Davis, *The Denver Post*)

Glenn Jones, who pushed the power of education through technology, was one of the many members of the cable gang who cashed out their companies as the industry consolidated into conglomerates in 1998 and 1999. (Karl Gehring, *The Denver Post*)

Comcast president Brian Roberts *(left)* struck a coup in June of 1997 when the world's richest man, Microsoft's Bill Gates *(right)*, invested $1 billion in Comcast as a show of confidence in cable's ability to offer high-speed Internet access. (AP/Wide World Photos)

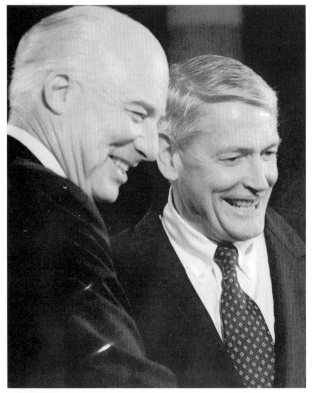

Endgame for the chess machine. TCI's John Malone *(right)* announces the sale of TCI for $48 billion to AT&T and its chief, C. Michael Armstrong *(left)*, in June of 1998 in New York. With a net worth of $4 billion, Malone continued to manage the Liberty Media programming and technology portfolio. (AP/Wide World Photos)

KLINE: God and the FCC willing.

MALONE: Yeah, well if it helps, I'll make a commitment to [the vice president], OK? Listen, Al, I know you haven't asked for it, but we'll make a commitment to complete the job by the end of '96. All we need is a little help ... you know, shoot Hundt! Don't let him do any more damage, know what I'm saying?

KLINE: For the record, maybe you should say you're kidding.

MALONE: Not about getting the highway up by the end of '96.

The "shoot Hundt" remark was picked up by newspapers across the country. TCI issued a formal apology and Malone called Hundt to explain. Hundt said little publicly about the incident. "I think he thought it was pretty funny," recalled Chong.

The FCC was a cauldron of conflicts. Hundt and Quello clashed repeatedly, particularly over a mandate that TV broadcasters carry three hours a week of children's programming. Hundt sponsored it. Quello thought the proposal, and much of Hundt's regulatory agenda, were folly. Quello staffers set up a mock shrine to Hundt in their office. It consisted of a painting of Hundt—commissioned by a trade magazine for a story—under which was set a votive candle, incense, a Pinocchio doll with a long nose and an offering plate. The plate typically contained a half stick of gum, M&Ms and a couple of pennies. Hundt ignored the shrine, but bowed down to a much larger plate with bigger pennies.

Before Hundt's arrival at the FCC, Congress authorized the agency to auction airspace for new communications services. The money would go to the U.S. Treasury. Hundt seized the auctions as a personal crusade. He prized a photograph in which he stood with Clinton and Gore, holding an oversized check for $7,736,020,384 paid to the order of "The American Taxpayer." The money came from the auction of 99 wireless phone licenses. Billions more would come from other spectrum auctions. "The FCC has become the federal cash cow," said Hundt, whose face sometimes scrunched up in a self-satisfied grin. "We happen to be the biggest profit center of any government agency in history and I just love bragging about that."

Hundt claimed that the auctions were "fast, fair and efficient." They were also subject to abuse. Some successful bidders defaulted on their payments. Others faced allegations of bid rigging. The legitimate bidders who prevailed would, under most economic theories, pass on the costs to consumers. The auction revenues were a prepaid communications tax.

Meanwhile, the politically connected broadcasters got their digital spectrum for free for the stated purpose of making the transition to digital TV. Republican senator John McCain of Arizona termed it "one of the greatest scams in American history."

Air could be worth a great deal. Or very little. It all depended on the pitch.

Hundt supported an auction, or a trade-off by which broadcasters would offer free airtime to political candidates. The decision not to auction the digital TV spectrum was approved by the Republican-controlled Congress and the Clinton administration. They covered up the scam by claiming that the broadcasters' analog channels would be returned to the government and auctioned off for other communications services once the transition to digital TV was complete. Then they cooked the books to anticipate $21 billion in auction revenues to help balance the current federal budget.

HUNDT HAD HIS hands full his first two years at the FCC, particularly since Congress was pursuing a massive overhaul of the nation's 1934 Telecommunications Act. He was besieged by lobbyists and corporate executives seeking his counsel and favor on a wide range of issues.

Among Hundt's actions at the FCC was to appoint Scott Blake Harris to head the newly created International Bureau, overseeing satellite matters. Harris was a Harvard Law graduate who had been chief counsel for the Bureau of Export Administration in the U.S. Department of Commerce. He also worked with Hundt on Gore's presidential campaigns. One of Harris's jobs was to streamline approvals for domestic satellite TV service. The bureau, said Harris, "would never slow the growth of this industry."

One of the first matters Harris faced was the application by Dan Garner's Advanced Communications for an extension to construct and launch a satellite to the 110 DBS slot. This was necessary before Garner could apply to transfer the 110 license to Malone's Tempo and cash out.

In years past, the FCC granted all sorts of extensions and waivers to get DBS service launched. USSB got a two-year permit extension to align with DirecTV. A company called DirectSat had its orbital licenses transferred to EchoStar.

Not this time.

In April of 1995, Harris denied Advanced's application for a four-year extension, canceled its construction permit and reclaimed the 110 slot "for reassignment to others." The reason given was that, "after more than a decade, Advanced had not provided, and wasn't close to providing, DBS service."

Given the fractious history between Malone and Gore, it looked like a fix by Gore's minion at the FCC, Hundt, and by Hundt's appointee, Harris.

Harris said it was strictly by the book. "I simply looked at the commission rules and concluded that the permit extension wasn't allowed," Harris said in a 1998 interview. "I made the decision without telling Hundt. No one outside the international bureau was involved in or consulted about the initial decision in any way."

It set off a mad scramble. Malone and Primestar had waited too long to move on 110, thus giving the FCC a procedural excuse to take the license away from Garner and block the transfer. What they needed were friends in high places. But Malone, who knocked government at every turn, had few friends in the back-scratching environs of the nation's capital.

"The failure to get the 110 slot from Advanced was huge," said Robert Thomson, TCI's public policy chief at the time.

Attorneys for TCI/Tempo and Advanced immediately filed an

appeal to the full FCC board. EchoStar and DirecTV piled on the objections. EchoStar described TCI as the "archenemy" of DBS. DirecTV claimed that "Primestar is not today and never will be a competitive service to cable television."

Both EchoStar and DirecTV hoped that the FCC would simply divvy up the 110 slot between their two companies. They had another angle, as well: The longer the 110 slot sat unused by a rival, the less competition they faced. Of course, that was also true of Malone and the cable gang: The 110 slot was as important for what it could block as for what it could provide—competition.

A foolproof way to keep the other team from scoring is to hold the ball.

MCI officials began lobbying the FCC and Congress to auction the 110 slot—the last great space on the U.S. satellite TV frontier. Meanwhile, Dan Garner was in no-man's land: The FCC had taken his license back and his $50 million deal with TCI's Tempo was in limbo.

Garner tapped his Arkansas connections, getting the state's six-member congressional delegation to write letters to the FCC on his behalf, urging the return of the 110 license to Advanced. Betsey Wright, the former Clinton chief of staff in Arkansas credited with coining the term "bimbo eruptions" in the 1992 presidential campaign, had become a lobbyist with the Wexler Group. Garner hired her to lobby Hundt. She recalled Garner's early advocacy of DBS in Arkansas, even though she and Garner had no dealings when she worked in the governor's office.

"As Chief of Staff to Governor Clinton, it was tough to decide whether Garner was a nut—or a visionary," Wright wrote to Hundt. She went on to criticize the auction proposal as property seizure. "The government of our country does not take away the successful and valuable work products of its citizens just because it can make money by selling that work to others."

The lobbying on the other side was just as fierce. Senator Byron Dorgan, a Democrat from North Dakota—where he once called the satellite dish the state flower—and Senator McCain sent a letter to their colleagues urging an auction. It would, they said, "stop a sweetheart deal that would permit 60 percent of the cable industry [to] control spectrum which should be used for a competitor."

By a 98–0 vote, the Senate passed a resolution that authorized the FCC to auction the 110 orbital slot and a secondary slot at 148 degrees that had also been taken back from Advanced.

"The fact is, this senator is very anxious, like all senators, to find money," Senator Ernest Hollings, a Democrat from South Carolina, said during the floor debate. Before the vote, an amendment was slipped into the bill by Republican senator Hank Brown of Colorado, at the urging of TCI's Thomson. It gave TCI wiggle room to make an offer to the FCC to pay the government "fair market value" for the 110 slot—however that was determined—without an auction.

But MCI already had an inside track.

In early October of 1995, Hundt and MCI's Bert Roberts, among hundreds of other officials, attended the International Telecommunications Union conference in Geneva, Switzerland. The two men discussed the DBS matter, Hundt acknowledged later. News of their encounter sent tongues wagging. "Would this same FCC chairman like to be the one to deliver his Congressional overseers many millions of corporate cash for the U.S. Treasury?" asked *DBS Digest.*

Soon after the Geneva meeting, MCI president Gerald Taylor sent Hundt a blunt letter, detailing the company's position. Copies were sent to the four other FCC commissioners as well. "MCI reaffirms its commitment to participate in the auction and will submit an opening bid of $175 million," it read.

Within a week, the FCC's decision was announced. The tally was 3–2 in favor of auction. Hundt cast the deciding ballot, joining commissioners Chong and Susan Ness, a Democrat. Hundt did not recuse himself, even though his old law firm, Latham and Watkins, represented DirecTV in its filings before the FCC.

The next day, Hundt pounded his chest about the virtue of auctions and shed crocodile tears for Garner. "If poor Mr. Garner had to buy his spectrum at an auction 10 years ago, he never would have been in this fix to begin with," Hundt told *Satellite Business News*. Asked if there would be DBS service at the 110 slot a year hence, Hundt said, "A year from now the market will have answered this instead of the FCC and that's the way things ought to be."

McCain also approved. "For once, the millionaires didn't get their way," he said, referring to the cable gang. Yet it would just be a different set of millionaires, MCI and Murdoch's News Corp., that would get their way.

Left to eat trail dust was Dan Garner.

FCC commissioners Quello and Andrew Barrett, a Republican, voted against the auction ruling. "The majority gives companies that chose to sit out the hard development days of DBS a windfall chance to participate in a gold rush, and leaves one of the pioneers of the DBS service with only a panful of mica," wrote Quello, referring to Garner.

Three years later, Garner attributed his failure to become king— or at least wealthy court jester—of satellite TV to his own bungled business decisions, poor timing and a political fix aimed at Malone. "If you get in bed with TCI, and there's any regulatory things in Washington, you're going to get crushed," he said.

The Consumer Federation of America, among other groups, approved the taking of the 110 slot from Malone's hands. Yet, TCI/Tempo was allowed to bid on the slot when the auction occurred three months later. Malone was going to get another shot. Then everyone would know how much the chess machine really thought 110 was worth. Many companies were expected to participate in the auction. But the Baby Bell phone companies didn't show up. Neither did DirecTV or its partner, USSB.

The day before bidding began, AT&T announced it would spend $137.5 million for 2.5 percent of DirecTV, with the intention of selling satellite TV service to its phone customers. "If we had opted to bid, it would have cost us $1 billion and two years to get to market," explained Joseph Nacchio, a top AT&T executive at the time. As it turned out, it would cost more than two-thirds of $1 billion just to claim the license for 110—a license that Dan Garner had once fruitlessly shopped around the halls of corporate America.

Three parties showed up to bid: TCI, EchoStar and MCI, with partner Rupert Murdoch in the wings. The auction began at 9:30 A.M., January 24, 1996, at the FCC's auction facility near Union Station in Washington, D.C. Privacy booths were set up for each of the three bidders.

MCI had a crush of people in the room, with a bus set up outside for executives to communicate by phone with Roberts and Murdoch. David Beddow represented Malone and TCI/Tempo. Ergen handled the bidding for EchoStar. Barely known inside the room, Ergen was a nonentity to the business world at large. That would soon change.

Each of the three bidders could have captured the 110 slot for a fraction of what was about to be paid: Ergen, if he had closed a deal with Garner; Roberts, if he had responded to Garner's 1993 letter; Malone, if he and Primestar had moved faster.

Garner stood in the back of the room. He wanted to participate in the auction, but couldn't raise the $10 million entry bond. Ergen came up to him at times to discuss the bidding. Someone introduced Garner to Scott Blake Harris. They shook hands, but had little to say to each other. "The American consumer will be the real winner of this auction," Harris told reporters. "It will lead to the prompt introduction of new DBS competitors and additional DBS service."

The minimum opening bid was set at $125 million. Each participant bid without knowing what the others offered until the end of the round. Each succeeding round required a bid at least 5 percent higher than the previous highest amount to continue the auction. The bidding was done through computer screens, which asked for this confirmation: "Are you sure you really want to do this?"

TCI broke out of the gate in the first round, bidding $201 million. MCI came up with $175.2 million. Ergen bet the minimum, $125 million. After 11 rounds, TCI dropped out as the bidding approached $300 million. "We weren't going to bid just anything," said Beddow.

At the end of the day, but not the end of the auction, MCI led the bidding at an astonishing $450 million. "There is no business model that anyone has seen to spend that much money, with three years getting to market, and still make a profit," said Bob Scherman, the editor and publisher of *Satellite Business News*, handicapping the action afterward. "I think TCI and EchoStar bid it up and MCI went for the bait."

It was Ergen's move.

Sixteen years earlier, he had been kicked out of casinos for card counting. At Reed Hundt's casino, with hundreds of millions of dollars on the line, Ergen hung in with MCI round after round. "We felt it was worth $650 million," he said later. "If we got the slot, we had a satellite under construction that we could have used it for. If we didn't, we raised the price of poker for somebody else and made our spectrum that much more valuable. So, it was one of the few times, at least since I've been in business, that I had a no-lose strategy."

It was not that tidy at the time. Behind the scenes, EchoStar executive Carl Vogel was scrambling like mad to get financing in place in the event that Ergen actually won. In the eighteenth round, Ergen bid $650 million. MCI beat it with $682.5 million. Ergen folded with a smile on his face.

"We went into this auction with the intent to win," said Roberts afterward, announcing MCI's venture with Murdoch to start American Sky Broadcasting, which would likely require another $2 billion to get off the ground.

"I think we're moving to an increasingly wireless world," explained Murdoch, expecting his third time pursuing DBS in the United States to be a charm.

As Ergen suspected, the wave of money spent for the 110 license buoyed the entire DBS industry. The Greater Fool theory was at play. "If a license, a piece of paper, is worth $682.5 million, then extrapolate that into the potential DBS market," said Peter Aseritis, an analyst with CS First Boston in New York.

A day later, EchoStar bid $52.3 million to beat out MCI for the 148-degree orbital slot, which covered mainly the western United States. EchoStar's stock shot up 27 percent.

The auctions confirmed what Dan Garner had felt in his bones for a decade: DBS was worth big bucks. But he wouldn't be seeing any of it. Just after the auction, reports surfaced that Malone was talking to Murdoch about joining the MCI/News Corp. venture. Malone had two Loral satellites on the ground, but nowhere to put them in space. "All I can say is, in a den of thieves, it's hard to be surprised," Garner told the *Arkansas Democrat-Gazette*.

Three weeks after the DBS auction, Reed Hundt gave a speech to the Artists Rights Foundation digital technology symposium in Los Angeles, where he fancied himself a telecommunications Robin Hood:

> I always say our auctions are not about the money. They're fast, fair and efficient ways to allocate precious public resources and jumpstart competition. It really doesn't matter how much we raise in the auctions. What is really important is that the government doesn't pick the winners of the licenses according to who has the best lobbyists or the most congressional influence. And we don't tell the license winners what to do to make money in the market-place. We guarantee that the markets are competitive and then we let the auction winners compete, whether it's in the cellular telephone business or the direct broadcast satellite business.
>
> So that's why it wasn't really relevant that Rupert Murdoch had to give me a king's ransom, an absolute fortune, a colossal sum, $682.5 million in an auction for the DBS satellite slot.
>
> The amount didn't matter at all.
>
> Who am I kidding?
>
> I can now reveal the truth about the auctions. Don't pass it on. We intended all along to raise in auctions of the airwaves more money than Bill Gates has.

Angling to get into the DBS game by any means possible after the 110 debacle, Malone proposed delivering service to the United States through a partnership with Telesat Ventures, a Canadian company. Telesat had a Canadian orbital slot. Malone had satellites.

Once again, Malone was hammered in Washington. The U.S. Departments of State, Justice and Commerce, plus the Office of the U.S. Trade Representative, wrote a joint letter on July 1, 1996, to the FCC opposing the plan. They cited a long-standing cultural trade dispute with Canada, then noted, "There is some concern that TCI, given its market position, may have an incentive to engage in anticompetitive behavior in parts of the U.S. market."

Scott Blake Harris, who by that time had left the FCC to set up a communications law practice in Washington, commented in a news story, "It is something that happens only rarely. The executive branch is sending a very clear signal."

The same could be said for Harris's initial decision on 110.

In private practice, with a Washington law firm named Harris, Wiltshire & Grannis, Harris's clients included Rupert Murdoch's News Corp. On his law firm's website, Harris touted his government regulatory experience and posted an article titled "New DBS Rules: From the Sublime to the Political." It criticized proposed FCC rules to prohibit cable companies from owning DBS licenses. "Nothing can stifle the growth of an industry more than the imposition of rules on who may invest and participate," he wrote.

Late in 1996, Murdoch and Roberts picked up shovels to break ground on ASkyB's $100 million satellite uplink in Gilbert, Arizona, the home state of Senator John McCain. Arizona beat out Colorado Springs, where MCI already had a major facility, after the Arizona state legislature offered tax breaks.

The wild card in all of this was Ergen. He was cool at auction, forcing Malone to fold and the Roberts/Murdoch team to pay through the nose. "We went in prepared and very nonemotional," Ergen recalled. "I felt like I'd been practicing for that particular moment for half my life."

NINE

ELVIS IS IN
THE BUILDING

T HE PLAYER TO WATCH IS the one betting with chips he can afford to lose. By the time he was 40 years old, Ergen had drawn $15 million from his $200 million big-dish satellite business to support his family in comfort, including a home in the Denver area and a ranch in southwestern Colorado. His decision to go into the small-dish business would put his company more than $1 billion in debt.

The real risk, as Ergen saw it, was standing still.

By the early 1990s, the big-dish market was stagnant, soon to be overtaken by Primestar, DirecTV and whatever other company could build and sell a better rat zapper. Ergen was game. In 1992, the government awarded him a DBS license at the 119 slot, consisting of just 11 channels. He immediately signed a contract for seven satellites.

It was all seven or Chapter 11, according to Ergen. "In business school they tell you not to put all your eggs in one basket. But I say put every damn egg you've got into it. If satellite TV isn't successful, we'll go out of business." If the company figuratively burned down, Ergen told associates, he'd toss the keys into the blaze and walk away.

That kind of talk provoked Malone who could counter any rational strategy but was unnerved by an adrenaline junkie like Ergen who claimed to have nothing to lose.

Ergen tested the small-dish business by becoming one of five national distributors for DirecTV. But that relationship soured when Ergen rallied his network of independent big-dish dealers around his own DBS plans.

He regrouped his privately held Echosphere companies into EchoStar Communications Corp. in 1994. Carl Vogel, a veteran programming chief at Jones Intercable, was hired as EchoStar's chief operating officer. "I thought, 'If these guys are half-right, this is a hell of a play,'" Vogel recalled in 1998.

EchoStar issued $335 million in junk bond debt, paying a whopping interest rate of nearly 13 percent. Investors snapped it up. The payments were backloaded so that EchoStar would not begin paying interest and principal until 1999. That gave Ergen five years to play. His mantra was to sell dishes faster, better and cheaper. "Everybody knows what we're about," he said. "We're running off-tackle to the right. Try and stop us."

EchoStar built a $40 million satellite uplink center on a 50-acre site outside Cheyenne, Wyoming, 100 miles north of Denver. Ergen picked the site for three reasons: cheap land, low taxes and reasonable electricity rates. One day, Ergen and Vogel drove up to take a look. Cattle grazing on the open field took an intense interest in their red automobile. "We better get out of here," said Vogel.

EchoStar went public in 1995, selling four million shares at $17 apiece under the ticker symbol, DISH, which stood for Digital Sky Highway. Going public meant that EchoStar's books were open. Ergen and his wife, Candy, owned 75 percent of the company's total stock value and more than 90 percent voting control.

EchoStar was a tugboat among supertankers, overshadowed in market power and financial muscle by companies like General Motors' DirecTV, Rupert Murdoch's News Corp. and Malone's TCI. Yet, the tugboat had at least one advantage over the supertankers: maneuverability.

Even Murdoch and Malone did not possess Ergen's freedom. He controlled so much of EchoStar's stock that he could act with impunity. Casually dressed among the suits of corporate America and policy makers in Washington, Ergen sold only one message, which was satellite TV versus the cable gang. He endlessly lobbied Congress to allow satellite TV firms to deliver broadcast channels into local markets. He became a poster boy for competition. Even when his stockholdings launched him onto the *Forbes* list of the 400 richest Americans in 1997, Ergen spoke of himself as the little guy fighting "General Motors and General Malone."

Throughout the 1990s, he acquired DBS licenses from companies including Direct Broadcast Satellite Corp. and DirectSat, which had yet to launch satellites. Ergen failed to nab the 110 slot from Dan Garner, but he did pay $52.3 million for a partial slot at auction.

There were eight orbital slots, with 32 transponders—or channels—at each position, or 256 channels total. Ergen controlled 90 of them, more than anyone else. With digital compression, that could mean 900 channels. It was a real estate grab in outer space that no other player attempted and few claimed to understand. "I just look at what's the cheapest way I can get information to your house," said Ergen. "The cheapest way I can do that is through DBS."

It was risky, but not mysterious. It looked much like Malone's strategy a decade earlier of buying up almost every cable system that came on the market. Ergen learned a great deal by watching the chess machine. EchoStar's supervoting shares and cash flow valuations could have come from Malone's playbook. The two men once met to discuss a potential merger of EchoStar with the cable gang's Primestar. Ergen wanted control of the joint company. Malone wanted to keep competition away from cable—or at least away from TCI.

"The gist of the meeting with Malone was: Here is what is in everybody's best interest," Ergen recalled. "They'd sell Primestar. We'd sell EchoStar. The TCI cable territories would be kind of off-limits." But Malone and Ergen could not come to terms, and EchoStar became an everlasting thorn in TCI's hide.

EchoStar's first step was to launch a DBS bird, a process Ergen admittedly knew little about. He relied on his satellite vendor— Lockheed Martin, after a series of mergers—for engineering, construction and $66 million in loans. To get service up quickly and cheaply, Ergen took another huge risk. He chose the China Great Wall Industry Corp. and its Long March rocket to deliver EchoStar's first satellite into orbit from Xichang, China.

After the Chinese government's murderous Tiananmen Square crackdown on pro-democracy demonstrators in 1989, such technology transfers required an export waiver from the U.S. government. EchoStar could have run into problems on this front. Operating as Echosphere, the company had settled charges in 1991 of export violations involving General Instrument decoders.

By 1994, however, the Clinton administration had declared the United States open for business overseas. Commerce secretary Ron Brown globe-trotted to the world's capitals with corporate chiefs and campaign donors in tow looking for deals. Ergen wasn't invited on one of those trips, but neither did EchoStar have a problem getting an export waiver for the China launch. "It is in the national interest of the United States to waive restrictions as those restrictions pertain to the EchoStar DBS project," according to Clinton's report to Congress on July 13, 1994.

EchoStar's contract called for China Great Wall Industry to launch not only EchoStar's first satellite, but its second and third as well. Ergen was courting disaster.

The Long March rocket was named for the 6,000-mile retreat that Mao Zedong and the Communist army made across China in the mid-1930s. The record of the Long March rocket lifting satellites into orbit was equally excruciating. Three of six Long March 2E launches prior to EchoStar's had failed, including two explosions shortly after liftoff. This was far worse than the industry's 15 percent failure rate. EchoStar's satellite and rocket were insured for the $200 million cost. But a launch failure would cripple the company's business plan and raise sharp questions about Ergen's judgment.

Ergen, his wife, and a small group of EchoStar officials took the 20-hour flight from the United States to China to witness the launch on December 28, 1995. About 400 EchoStar employees, friends and family members gathered before dawn at company headquarters south of Denver to watch via satellite. Ergen had sunk the company deep in debt for a chance to enter the DBS business. His future rode on a Chinese firecracker as likely to blow up as go up. Success was so uncertain that EchoStar prepared a press release anticipating the worst. "Obviously, we're very disappointed with this morning's launch," was the canned quote from Ergen, above a line that read "Additional information on the launch failure will be released by EchoStar pending results of a full investigation."

It wasn't necessary. The EchoStar bird with the 120-foot wing-span was successfully launched to the 119 orbital slot. A 15-year journey from crashing a satellite dish by the side of the road to launching a satellite one-tenth the distance to the moon was complete. From then on, when anyone questioned EchoStar's strategy or doubted Ergen's ability to pull it off, he had a ready comeback. "I bet this company on the nose of a Chinese rocket!" he'd say, hardly believing it himself. "A Chinese rocket!"

It was beyond luck.

ON FEBRUARY 15, 1996, six weeks after the successful launch of *EchoStar I*, a Long March 3B rocket carrying a Loral satellite exploded after its launch in China. In a nearby village, several people were killed and dozens injured from falling debris and burning fuel. The Intelsat satellite project was to serve Latin America with satellite TV service, backed by a consortium that included Murdoch's News Corp. and the international arm of Malone's TCI.

The launch failure sent reverberations far beyond that. Loral and GM Hughes Electronics were investigated by a federal grand jury and Congress for helping Chinese engineers improve Great Wall's launch capabilities, particularly after the 1996 explosion. The security concern was the crossover between commercial and military

uses. Besides launching private satellites, Great Wall tested and provided technology for China's nuclear weapons. Great Wall was
sanctioned by the U.S. State Department in 1991 and 1993 for selling missiles to Pakistan.

The Clinton/Gore administration faced Republican allegations
in 1997 that it had compromised national security by turning a
blind eye to sensitive technology transfers to China, while accepting
donations from vendors and Chinese sources to fund their successful 1996 reelection campaign. Loral chairman Bernard Schwartz
was the single biggest donor to Democratic campaign efforts, with
contributions exceeding $1 million.

Meanwhile, GM Hughes chief C. Michael Armstrong, head of
Clinton's export council, had been among those pressing for quicker
approvals of satellite sales overseas to compete with vendors in other
countries. In March of 1996, shortly after the rocket explosion in
China, the Clinton administration transferred satellite export li-
censing from the Pentagon to the Commerce Department. Three
years later, under intense scrutiny, the administration blocked
Hughes from selling a $450 million satellite project to Asia-Pacific
Mobile Telecommunications because of concerns the technology
could advance the Chinese military's communications structure.
The Cox Committee report, which raised disturbing questions
about national security breakdowns involving China during the
Clinton administration's watch, was released in May of 1999.

Loral and Hughes denied any wrongdoing.

While shards of the full story surfaced, Ergen had already dodged
the fallout. *EchoStar I* was in orbit and operating. The 1996 launch
explosion allowed Ergen to sever his long-term contract with China
Great Wall Industry Corp., which he did. EchoStar's next three
launches were with French-based Arianespace, Lockheed Martin in
the United States, and International Launch Services, a partnership
between Lockheed Martin and Krunichev of Russia.

Had his Chinese launch failed, Ergen would have had little credibility to bid $650 million for the 110 slot in January of 1996 against
Murdoch and Malone. Had the launch failed, EchoStar could not

have rolled out its 100-channel Dish Network satellite TV service in March of 1996. "Had the launch failed, I think we might have gone out of business," Ergen said in 1999.

It was that willingness to dance on the edge, along with his boyish enthusiasm, Tennessee twang and loyal following among EchoStar dealers that led some within EchoStar to refer to Ergen as "Elvis." But the name also reflected a darker shade of Ergen's character: He could be a control freak with a mean streak.

"I heard the f-word as an adjective, a verb, a noun and a dangling participle," said one former EchoStar executive. "He screamed and yelled 20 times a day."

Amid Colorado's laid-back business culture, full of taciturn natives and rat-race exiles from both coasts, EchoStar resembled a Silicon Valley start-up in perpetual crisis. The culture was workaholic: If you don't come in on Saturday, don't bother coming in on Sunday. Either day was likely to find Ergen at headquarters—on the phone, answering mail or prowling the company's technology lab, where EchoStar engineers devised state-of-the-art consumer electronics. Ergen, a CPA, signed every company check in the early days, then every check over $10,000 later on. "That way, I know everything that goes on," he said. "It's the best flow of information."

During the years building Echosphere, Ergen had cultivated a network of hundreds of independent dealers who didn't eat if they didn't earn commissions from selling and installing satellite TV systems. Ergen traded e-mails with them and hosted "Charlie Chat," a monthly closed-circuit video conference to discuss promotions and pricing plans. He gained detailed knowledge of field operations, like a general receiving updates from the front lines. Ergen devoured this information and sometimes used it to test his executives. "What's going on in the field?" he'd casually ask, already knowing the answers. Later, when Ergen hosted a monthly call-in show for EchoStar subscribers, also called "Charlie Chat," company officials watched anxiously as Ergen ad-libbed new program packages or pricing plans on the air, which they would then have to implement the next day.

EchoStar's focus group, they realized, was when Charlie looked in the mirror.

The intensity took its toll at work and at home. Candy Ergen called EchoStar her "sixth child." Yet, raising their five children while her husband raised the sixth wore thin at times. "I want my husband back," she told one EchoStar official. "He gives you a goal and just as you get to the goal line, he moves it."

Ergen's nervous energy permeated the building. An assistant to Carl Vogel alternated red and green screen savers on her computer to warn employees whether it was safe to approach Vogel's office. It was often red. Vogel negotiated programming contracts for Echo-Star and found himself shifting money like a street-corner dealer hustling three-card monte. One crunch came when EchoStar needed a deposit for its second satellite launch. "I remember doing a deal with QVC, just so they would pay us $5 million up front for carriage," Vogel recalled. "We took the check and immediately wired it in to Arianespace to reserve the launch."

Vogel described Ergen as "driven, focused, a little crazy. He challenged me more than anyone I've ever worked for. He never pushed me too hard, but I saw him push others too hard."

A famous line from Elvis Presley's later years was "Elvis has left the building"—an announcement meant to disperse the adoring crowds. At EchoStar, Ergen rarely left the building, but a string of executives did—for good. Four officials, including three vice presidents, quit for various reasons in one week in 1996. Much of EchoStar's top management team remained in place from the early days. They were inured, if not immune, to Ergen's tirades. Their salaries were typically less than at major media companies, but there was the promise of future wealth through stock options if EchoStar succeeded. Ergen prided himself on cutting costs to the bone. EchoStar executives double-bunked on the road and were encouraged to travel by red-eye flights so they didn't miss a day's work. Ergen himself struck a deal with United Airlines, paying for a lifetime, best-available seat on any domestic flight. EchoStar bought national newspaper and magazine ads at 80 percent discounts at the last minute rather than pay full price in advance. Cost-cutting ruled the day. There were heated debates at EchoStar about how much to pay for barbecue supplies at dealer promotional events.

In Ergen's eyes, EchoStar was a laser focused on just one target: selling more satellite dishes. No gimmick was left untried. An actor was hired and dispatched to the Consumer Electronics Show in Orlando and other events dressed as "Dishman, the fearless crusader for better television." An easy mark for newspaper photographers, Dishman had programmer logos plastered on his jumpsuit and a satellite dish attached to a helmet on his head. EchoStar even tried to wring a testimonial from country singer Willie Nelson after installing a dish at his home in Austin, Texas.

Ameristar, an independent satellite TV distributor based in Nashville, held rallies in hotel ballrooms across the country, luring potential salespeople with the chant, "Cable isn't able! Cable isn't able!" Ameristar officials claimed that by selling and recruiting others to sell—a practice known as "network marketing" when it succeeds and a "pyramid scheme" when it fails—an aggressive vendor of Dish Network equipment could earn $1.5 million a year. That was seven times Ergen's salary. One Ameristar tip for selling high-tech satellite gear was decidedly low-tech: A handout suggested that salespeople use a spoon or a pen to imprint customer credit card numbers on carbon copies when closing a sale.

Ergen's secret weapon, however, wasn't a gimmick. It was the fact that the company manufactured its own equipment at plants in Huntsville, Alabama, and overseas. Ergen knew what a dish, a set-top box and a remote control cost down to the penny, and he knew he could undercut the competition.

EchoStar began selling Dish Network hardware in 1996 at $499, plus installation and monthly programming fees. They captured about 50,000 subscribers, a fraction of the one million apiece claimed by DirecTV and Primestar. In June, EchoStar dropped the hardware price to $199 in seven markets where cable rates went up sharply. To get the deal, subscribers needed to prepay $300 for a year's programming of 40 channels. It was the same $499 up-front cost as before, but with programming included.

At first, DirecTV dismissed the move as an act of desperation. "We look at our service as the Cadillac of DBS service," sniffed Linda Brill of DirecTV. "If someone wants to go to Ford, then that's fine. We don't think EchoStar's program is going to last that long."

It not only lasted, but Ergen extended the offer nationwide. The company lost $300 in equipment and marketing costs for each system sold. They gambled on keeping customers long enough to earn that money back and much more in future years. It sounded like the punch line to an old joke: We lose money on every customer, but make it up on volume.

DirecTV president Eddy Hartenstein was not amused. It was too early in the game to sacrifice market share, so Hartenstein met with his company's equipment vendor, RCA-Thomson. "We said, 'We're Hughes. We're part of GM and we're in this for the long haul,'" Hartenstein recalled. "If that's what it's going to take, we can play that game, too. So, everybody took a bit of a haircut."

Perhaps $300 million in potential revenue was shorn from all players by Ergen's lowball pricing. DirecTV dropped its equipment price to $199 and even offered new customers in 18 cities a programming price freeze. Primestar subscribers, who paid $10 a month to lease their dish, could own a DirecTV or EchoStar system for less than they paid Primestar over two years. Ergen not only undercut his satellite rivals but took aim at the cable gang, particularly TCI, which raised rates as if it still had a monopoly—up to 20 percent in some markets in the summer of 1996. "I wouldn't say this is necessarily the death knell of cable, but the days of them owning the street are over," said Larry Smith, an EchoStar vice president.

Waking from its slumber, TCI ran print and radio ads describing the downside of small-dish satellite TV: the up-front cost, no broadcast channels, an extra charge to connect a second TV set and potential signal interference called rain fade. The ads ended with the tag line, "TCI. We're a little more down to earth." EchoStar responded with full-page newspaper ads signed by Vogel: "Since when has the same company that raised its rates over 20 percent a few short months ago become your consumer advocate?"

The advertising volleys were refreshing bursts of competitive fire, proving that the viewing dollar of the country's couch potatoes was actually worth fighting for—not simply a monthly annuity paid to the local cable monopoly and its program suppliers.

EchoStar kept up the sniper attacks. When TCI lost a public vote on its cable franchise renewal in Boulder, Colorado, by a two-to-one margin in 1996, EchoStar gave away satellite dishes in the college town, which is better known for rock climbing than channel surfing. A nationwide "Cablebuster" promotion cut $100 off EchoStar's $199 hardware price if buyers showed a current cable bill.

In September of 1996, EchoStar launched its second satellite, on an Arianespace rocket from French Guiana. It doubled EchoStar's channel choices to 200. Ergen engaged in some hagiography before the launch, telling a reporter, "If this were the Old West, I'd be the guy in the covered wagon, the explorer." One of his employees, watching the *EchoStar II* launch at company headquarters, put it less romantically: "We're making it cheaper to watch Jenny Jones."

That was what worried the cable gang. With 65 million subscribers, they still dominated the landscape. Wireless cable companies and other competitors had only a handful of subscribers in comparison. Yet, a Yankee Group study projected satellite TV capturing 15 million homes by year-end 2000. That was 15 million that cable would either lose or not win going forward. And the first cable customers to embrace DBS were ones with annual incomes greater than $75,000.

Cable's growth prospects soured and stock prices fell. As if by kismet, a dark comedy appeared in movie theaters in the summer of 1996, titled *The Cable Guy*. It starred Jim Carrey as a maniacal cable installer. Beyond a few odes to piracy like "Free cable is the ultimate aphrodisiac," the movie itself had little to do with the business of cable television.

But it jolted cable officials trying to mend the industry's image. National Cable Television Association president Decker Anstrom attempted humor in a speech in Nashville, Tennessee. "The first version was 'The Post Office Mailman,' who suddenly goes berserk and grabs a rifle, and then begins to kill people, but that story was judged to be too dark—and likely to be mistaken for a documentary," Anstrom told a meeting of the Society of Cable Television Engineers. He cited other scenarios involving the telephone and satellite TV industries, then said, "We're self-confident enough to laugh at this film not only because we know it's pure fantasy, but also because we feel fortified by today's reality: a fully rejuvenated cable industry."

But cable's troubled tale was about to open nationwide—for real.

TEN

ANGRY

EYEBALLS

I N JANUARY OF 1996 at TCI headquarters, John Malone addressed a group of employees and guests of Tele-Communications International Inc., the arm of TCI that managed investments outside the United States.

As was his habit, Malone spoke fluently, without notes:

> One of the issues that we've had to confront over the last six or seven years is: What does a company do when the technology upon which it is based starts to change radically? When the tectonic plates that you've built your empire on start to shift, how do you handle that? What's the attitude? Do you become defensive? Do you try to use regulatory or technical barriers to prevent unpreventable change or do you try to go with the flow?
>
> I think we decided to go with the flow at TCI. We decided we couldn't predict whether terrestrial or satellite or wireless was going to dominate distribution of communications—both one and two-way—so we decided to participate in them all.

Malone was like a roulette player who bet all over the table. TCI was the largest cable company in the nation and owned part of the largest one in England, TeleWest. It owned part of Primestar and 30 percent of Sprint PCS, a wireless phone alliance with Sprint, Comcast and Cox. TCI's Liberty Media held stakes in dozens of networks, including 9 percent of Time Warner and its programmers. Malone was also a founder and chairman of CableLabs, a research and development arm for the industry, based near Boulder.

TCI was at the intersection of all the action.

"Can we develop a business in which millions of users will pay a monthly service charge for access to our services?" Malone asked rhetorically. "That is the entire focus of everything we do, be it in programming, be it in technology, domestic or international. We believe that in the future that is the way to best create wealth for shareholders. Which is, after all, as a capitalistic enterprise, what our fundamental goal is."

As long as the cash flow serviced its debt, TCI could grow in every direction at once, seeking the killer applications—the new services that would convince people to spend their money. But by the mid-1990s, the promise of 500 channels and interactive television was trumped by the reality of interconnected computers—the Internet.

It seemed logical that Malone, head of the world's dominant cable company, and Bill Gates, head of the world's dominant software company, Microsoft, would find a way to dominate this new communications technology, either together or separately. Both men had the savvy to generate their own ideas and co-opt others. Gates and his team created the Microsoft disk operating system (MS-DOS) for IBM's fledgling personal computer division in the early 1980s. It became the standard, from which Microsoft launched a suite of programs and—parallel to the rise of the user-friendly Apple Macintosh—created the Windows software that ended up running in 90 percent of the world's personal computers, regardless of manufacturer. Malone, to a large degree, controlled the cable industry's operating system—that is, the pipes and the programming that flowed through it.

Like John D. Rockefeller in the oil business and Henry Ford with mass-produced automobiles, Malone and Gates shared a hyper-rational industrial view of the world and a lust for scale economics.

Physically, Malone was more commanding than Gates, like a broad-shouldered drill sergeant compared to a scrawny buck private. In their meetings, Malone could sit stone still while Gates nervously rocked his upper body back and forth. Yet, Gates was 14 years younger, many times richer and a cultural icon.

Where Malone was once celebrated as the visionary, the bell-wether for communications technology, his star in the public sphere dimmed considerably when the Bell Atlantic deal fell apart in 1994 and TCI's digital cable service was indefinitely delayed.

Meanwhile, Gates was hailed as a geek god, with bouquets—and some brickbats—tossed at his feet after the phenomenally successful launch of the Windows 95 operating system. His every action and utterance was scrutinized for signs of the future. He wrote a best-selling book, *The Road Ahead*, and was the subject of 10 others. Gates's image was used as a hook for TV news shows, magazine stories, newspaper articles, cartoons and Internet sites, one of which featured a wealth clock documenting the ballooning bankroll of the world's richest human.

Microsoft posted revenues no greater than TCI in 1996—about $8 billion. But Microsoft reported huge profits and was billed as an essential cog in the information economy. The jury was still out on TCI, whose stock market value was one-eighth that of Microsoft's.

Gates also succeeded Malone as the technomonopolist that official Washington loved to hate, the subject of ongoing antitrust probes. Their styles were different in response. Malone was a provocateur who spoke his mind, whereas Gates tried, with varying degrees of success, to soften the edges and play the game. When Microsoft first faced government antitrust scrutiny in 1994, Gates went golfing with President Clinton and others at the Farm Neck Golf Club on Martha's Vineyard. When photos appeared of Clinton driving Gates around in a golf cart, some wags observed, "The most powerful man in the world—and his driver."

Gates reportedly told Vice President Al Gore that if the government pushed too hard in its antitrust case, Microsoft could relocate out of the country and export its goods to the United States under the North American Free Trade Agreement.

"They've got to treat him a little bit carefully—he might bite," said Malone, referring to Gates. "Our problem is, all of our assets are here. They're hard to move. I can't threaten to move to Canada. Bill can."

With all the motivations for Malone and Gates to work together—that is, to expand their separate empires—their early partnerships were spectacularly unproductive. In 1994, TCI and Microsoft proposed a cable-TV network about personal computers that included a home shopping element. It never happened and the idea was later pursued by Ziff-Davis with ZDTV.

Then came America Online (AOL), the corner store in cyberspace. Founder Steve Case had a 1993 run-in with Gates, who reportedly told him, "I can buy 20 percent of you or I can buy all of you. Or I can go into this business myself and bury you."

Case was a huge fan of Malone and tried to get TCI interested in investing. Finally, in late 1994, they came to terms and Case faxed Malone a draft press release. Soon after, Malone announced that TCI was going to invest $125 million of TCI stock for 20 percent of Microsoft Network, Gates's fledgling Internet service. Case suspected he got used. "I had this vision of him sending off the rough draft of the press release to Gates the minute it came off the fax from me," he said.

AOL boomed and Microsoft Network floundered. By late 1996, Malone recovered TCI's stock investment. "It was a mistake" not investing in AOL, Malone said later. "But you don't know that at the time. We thought Microsoft would be more powerful in the Internet access service space, but Bill was kind of slow out of the blocks."

From its mainframe genesis in defense labs and university research links, the Internet had evolved into a new way to communicate—for those who were interested and could afford a personal computer, modem, software and phone line. Electronic bulletin boards sprang up, generating hopes that cyberspace could become a place for "community" and revitalization of democracy—a new utopia. On the commercial side, smut peddlers moved in to sell every imaginable fantasy and degradation. The undeniable hit was the most personal—e-mail. It had the immediacy of a phone call, the informality of a Post-it note and eventually dwarfed the U.S. Postal Service in number of messages sent.

Both Malone and Gates were slow to embrace the Internet. It was not until December 7, 1995—Pearl Harbor Day—that Gates announced a coherent Microsoft strategy for the Internet. Malone's attitude toward the Internet was "cultural discomfort," according to John Perry Barlow, an early Internet advocate who appeared with Malone on a panel in late 1994. Barlow, a Wyoming rancher, cofounder of the Electronic Frontier Foundation and sometime lyricist for the Grateful Dead, said he "laboriously built" for Malone "a credible business model selling Internet connections through the wasted bandwidth of his cable network." According to Barlow, Malone replied, "OK, you've convinced me there's probably a business to be made there, but it's not a business I want to be in."

Whether Malone was blowing Barlow off or really had his doubts, his interest in cable lines carrying Internet traffic intensified after a visit by John Doerr, the Silicon Valley venture capitalist. Doerr was part of the anti-Microsoft coalition that included Sun Microsystems, developer of the Java software, which could run with or without Microsoft's operating system. Sun's motto was "The network is the computer," implying that Microsoft's dominance was an unnecessary choke point. Some in the anti-Gates contingent supplied the U.S. Justice Department with evidence to make its antitrust case against Microsoft.

Doerr's connection to Malone was through Bruce Ravenel, a TCI technology executive. Doerr and Ravenel worked together in the 1970s at Intel Corp., the computer chip maker. Doerr pitched Malone four days before Christmas of 1994, on exactly what Barlow proposed. Within five months a company named At Home Corp. was formed, based in Redwood City, California. At Home's mission was to devise the communications backbone and service to carry Internet traffic at very high speeds back and forth through cable lines.

Telecom systems have organic analogies. The phone network is a wheel, with messages pouring down the spokes to a hub, then back out again. Satellites are suns, beaming video and data to vast areas. Cable systems are like trees laid underground, with the branches delivering one-way video to homes. Rebuilding cable systems with fiber-optic lines tied to coaxial cable and two-way amplifiers is intended to transform the tree into a wheel. Particularly since the sun is so bright.

"One-way cable TV is essentially dead," wrote deep thinker George Gilder in April of 1995. "In response to [satellite], cable has no choice but to change its business radically to two-way computer services."

With Ravenel as scout, Malone was moving TCI incrementally toward cyberspace. It bought 2.3 percent of Netscape, Microsoft's rival in the browser wars. When Netscape went public in August of 1995, its share price soared and helped set off the Internet stock craze. Malone later expressed mystification at the values of Internet start-ups with minimal revenues and track records. "It's great if you can capitalize a business like Netscape by having an idea, raising a shitload of money, then building a business," he said. "But it doesn't always work."

Malone also said he got "10 opportunities a week to invest in Internet content providers. They all have one thing in common: they're all going to lose $5 million this year." TCI's most promising investment was At Home, cited as @Home, the @ symbol having made the transition to the media mainstream by way of e-mail addresses.

Having learned his lesson from Primestar's convoluted ownership structure, Malone returned to what he knew best: control. Comcast and Cox became co-owners in @Home, but TCI retained majority ownership of the service, which would be offered to any cable company that wanted it. Time Warner pursued its own cable Internet service named Road Runner. MediaOne, the former Continental Cablevision company owned by US West, also went its own way.

Coaxial cable has much greater bandwidth, or capacity, than phone wires. @Home claimed it could download a video clip to a computer screen hundreds of times faster than possible through a traditional phone line and modem. This cable Internet solution fit perfectly with Bill Gates's concept of the "Web lifestyle," by which the worldwide computer network—and therefore, Microsoft—is integral to everyday life. "You're living a Web lifestyle when you just take it for granted that any purchase you make, any new thing you want to plan, like a trip, you turn to the Web as part of that process," said Gates. "People today live a phone lifestyle and a car lifestyle. And they almost laugh when you say that to them, because it's so taken for granted."

Gates wasn't laughing, however, when he found that Malone was @Home alone and would launch cable Internet service without ties to Microsoft. "Gates 'just exploded' at Malone as part of their scheduled one-on-one meeting, threatening to bury this company, buy cable operators, and do whatever it took to crush @Home, since we are obviously so anti-Microsoft that it's criminal," according to a May 1996 e-mail from @Home executive Milo Medin to Marc Andreessen, Netscape's founder.

The e-mail came out during the government's antitrust trial against Microsoft. It flipped a few assumptions. Malone, cable's enforcer, was in the crosshairs of a cutthroat who just looked benign.

⁓

BY MID-1996, Malone's troubles at TCI went much deeper than Bill Gates's threats. Malone's doctor told him in 1993 that his persistent flu symptoms were from stress. On a business trip to Argentina in early 1996, Malone caught a nasty bug. He spent a good deal of time at his vacation home in Maine and left TCI's day-to-day affairs to his lieutenants. Speculation swirled that the 55-year-old cable magnate was seriously ill, having a midlife crisis or had yet to rebound from the failed merger with Bell Atlantic.

Maybe he was spooked by the Unabomber, from whom Malone later claimed to have received a threat.

Theodore John Kaczynski opposed everything that technologists like Malone and Gates stood for—and he aimed to destroy it one human at a time. Working from an isolated cabin in Montana, Kaczynski crafted mail bombs that killed three people and seriously injured eight people between 1978 and April of 1995. The primary targets were university researchers and business executives.

In September of 1995, as part of a negotiation with the Unabomber intended to halt the bloodshed, *The Washington Post* and *The New York Times* jointly published the Unabomber's manifesto. It included these footnotes:

> We are asserting that ALL, or even most, bullies and ruthless competitors suffer from feelings of inferiority. ... The conservatives are just taking the average man for a sucker, exploiting his resentment of Big Government to promote the power of Big Business. ...
> When someone approves of the purpose for which propaganda is being used in a given case, he generally calls it "education" or applies to it some similar euphemism. But propaganda is propaganda regardless of the purpose for which it is used. ... Many leftists are motivated also by hostility, but the hostility probably results in part from a frustrated need for power.

A psychologist who examined Kaczynski later wrote that he had "an almost total absence of interpersonal relationships" and "delusional thinking involving being controlled by modern technology."

By the mid-1990s, those who considered themselves a potential target had beefed up security measures. At the home of Bob Magness, packages were x-rayed before they came into the home. TCI took similar precautions at headquarters. Malone already lived in the country, at the end of a winding driveway, behind a security gate.

In February of 1996, Kaczynski's brother, David, contacted the FBI with his belief that his brother might be the Unabomber. The FBI began its manhunt, which was increasingly publicized. On April 3, 1996, FBI agents arrested Kaczynski at his Montana cabin. They found bomb-making materials and a list of 400 people who were potential targets. Those people were contacted by authorities, but they were never named publicly. At his trial, Kaczynski was sentenced to life in prison without possibility of parole.

——

ON APRIL 2, 1996, Malone spoke at the fifth anniversary party for Encore Media, a TCI-controlled programmer. He spoke to reporters afterward about his reduced operating role at TCI. "Virtually all my wealth is tied up in the various companies," he said. "Obviously, I'm not going to lose interest. The problem is, some of us get old and we have to let others take over." By that time, Malone had cashed out his $10 million personal stake in TCI's international company, which sent that stock spiraling down.

When Malone caught cold, the cable business got pneumonia. Cable rate hikes averaged 8.5 percent in 1996, more than three times the rate of inflation. This boosted cash flow, but helped drive subscribers into the beckoning arms of satellite TV companies. Cable's debt pyramid was cracking and TCI had the farthest to fall. By mid-1996, the six members of the cable gang carried a debt load of $41 billion—one third of which belonged to TCI. Something had to give.

The head of TCI's cable operations since 1994 was Brendan Clouston, a native of Montreal with an MBA and a banking background. He worked for TCI in the 1980s and rejoined the company in 1991 when it took over United Artists Entertainment, a cable company, where Clouston had been chief financial officer.

"Brendan is basically the president of the company, though we don't call him that yet," Malone said in 1993.

Clouston was a strategic planner and number cruncher who moved in as J. C. Sparkman retired. Clouston lacked Sparkman's experience in cable field operations, but spent big on consultants and pie-in-the-sky technology. He acted as if TCI were actually going to build and operate all the things Malone talked about. Meanwhile, cash flow margins shrunk.

As the TCI supertanker steamed toward a financial iceberg, a presentation was held at company headquarters for the business press in mid-August of 1996. Leading it was Robert Thomson, TCI's gruff public policy chief. A graduate of the University of Washington, Thomson served as an infantry lieutenant in Panama and Vietnam before earning a law degree at Georgetown University. He was a congressional liaison in the Carter administration and worked as general counsel for a Connecticut transportation company before joining TCI in 1987.

Thomson knew his way around Congress and was TCI's chief flak catcher and spokesman. There were at least three views of Thomson, none of which were contradictory. One, that he magnified Malone's hubris. Two, that he was a loyal soldier. Three, that he was given few resources to do his job.

As reporters nibbled on scrambled eggs and toast and scribbled on pads, Thomson served up the latest spin. "This is going to be one of the best marketing companies in the country," he said. "I know that sounds grandiose. I know that sounds like utter bullshit. But when you look back three years from now, Brendan Clouston will have superintended the most striking transformation of any company in history."

Also in the room was Sadie Decker, Ph.D., a former aerospace executive who headed Summitrak, a $100 million TCI project to route operations and customer transactions through a supercomputing system. "The data center is sucking power, they're just generating data like crazy," said Decker. "I'm the one to catch the billions and billions of transactions."

Whatever image inflation occurred at the briefing quickly deflated when Thomson announced that no press would be allowed to attend the company's annual meeting the next week—an odd prohibition for a Fortune 500 company. Most public companies welcome the chance to get a staged event publicized in the business press.

But TCI was as prickly as they came, a corporate cactus.

The reporters discussed some leverage of their own: They could each buy a share of TCI stock at the stagnant price of $15, attend as cheapskate shareholders and risk getting kicked out.

"I'll sell you one," Gerald Gaines, the head of TCI's cable telephone division, said softly.

Thomson reversed the no-press policy the next day.

—

AFTER HE HAD BEEN diagnosed with cancer, but several weeks before he entered a Virginia hospital, Bob Magness sat for an interview in the rolling backyard of his home in Cherry Hills Village, the blue sky above and the craggy Rocky Mountains in the distance.

"Are you O.K.?" asked the interviewer, who worked for TCI.

"Sun's a little bright," said Magness, squinting, the folds of flesh gathering around his eyes like bread rising in the oven. Magness wore a robin's egg–blue, short-sleeved shirt with a wide collar. His arms were splayed at right angles on the arms of a wicker chair, as if he were ready to draw holstered guns. His thin white hair was parted neatly and combed to the side. A shy man, Magness was uncomfortable, even for this, a videotaped interview as part of a tribute to his friend and cable colleague, Carl Williams. Two months later, Williams would deliver a eulogy at Magness's funeral along with Malone and Ted Turner.

"I better get a little looser before I start talkin' a little easier," said Magness, nervously rubbing his left trigger finger against his thumb as the whoosh from a jet overhead drowned out the sound. "We did a lot of deals on a handshake," Magness said about dealing with Williams. "I'd as soon have his handshake as his contract. Sometimes we'd have to put it on a sheet or two of paper."

The interviewer attempted to draw Magness out, but the cable cowboy kept his mouth shut. The secrets, both good and bad, were locked away. "There's a lot of things I respect about Carl that they don't need to know," said Magness.

The interviewer suggested that Magness look into the camera, and even tried to put words in his mouth, but Magness was on his own. "I'm about wore out," he said.

—

IN A RAGING BULL market, cable stocks were dogs. Investors had many more promising technology and media companies to pick from. Intel and Microsoft were good bets. Internet stocks were the new gold rush.

The multiples of cash flow by which cable stocks traded were sinking. "We do not recall in more than two decades such a disconnect between stocks and fundamentals," wrote Dennis Leibowitz, a media analyst at Donaldson Lufkin & Jenrette. The drop "seems to represent sudden and final surrender by some of the industry's stalwart backers, since most of the disbelievers are long gone."

During TCI's annual meeting in August, Malone and Magness were quizzed by John Gilbert, a lifelong New Yorker and shareholder gadfly in his 80s who was also an honorary Ringling Brothers clown. He donned a red nose to greet Magness, asked the members of the press in the room to take a bow and handed Malone an 18-inch blue plastic shoehorn after the meeting.

"Now I won't have to bend over for the government anymore," said Malone.

It was funny, but TCI's financial situation was scary.

Its stock fell under $12, a 52-week low. TCI froze vendor shipments, which sent the stocks of those companies tumbling. Standard & Poor's put TCI bonds on Credit Watch, perilously close to a junk rating, which could raise the cost of borrowing and eventually force big institutional investors to sell off TCI stock.

An October 14 *Business Week* cover story titled "Cable TV: A Crisis Looms," painted a bleak picture of a $25 billion industry with a history of broken promises and splintered leadership that faced intense competition. Malone needed to make a statement. With Clouston in tow, he chose a Bear Stearns conference in Phoenix, Arizona, capping a two-day event that also featured Barry Diller and Seagram's Edgar Bronfman Jr.

In his dinner speech, Malone called the players in the small-dish satellite TV business "the seven dwarfs" and hyped TCI's long-promised digital cable service to provide more channels. "Rumors that I have expired or am terminally ill or have lost interest in the cable company are substantially inaccurate," he said. Malone liked the line so much, it was fed to reporters in New York at *The Wall Street Journal* and *USA Today* for the next day's newspapers.

Whatever the spin, Wall Street's passion for Malone's financial engineering had cooled considerably. "The issue is simple," said cable analyst John Tinker. "No one believes TCI is capable of introducing a new product."

"They haven't met any targets," Larry Haverty of State Street Research told the *Journal*. "It's like the emperor has no clothes."

More trouble lay ahead.

On November 15, 1996, Malone spoke to a Denver meeting of the Cato Institute, the libertarian think tank of which he is a board member. Malone lauded free enterprise and gossiped about the feud between Rupert Murdoch and Ted Turner over Fox News in New York. Malone harped on a favored theme: how government regulation and tax policy distorted the free market, the cable industry and TCI in particular. It was an unwitting epitaph for the battles he and Magness fought in order to build their empire over the years.

The next day, Magness died of cancer at age 72. He had been treated for fast-moving lymphoma at the University of Virginia Hospital in Charlottesville, a facility recommended to Malone by Michael Milken, who had survived prostate cancer and funded cancer research.

At Magness's bedside when he died were his two sons, his wife Sharon and Malone.

His 22-page will had handwritten notes and charitable bequests to the Boy Scouts and the University of Denver scrawled in the margins. The will was signed and witnessed nine months before his death, with most of the assets left to his two sons, Gary and Kim. However, the will failed to shelter the vast family fortune, mostly in TCI and Liberty stock, from a 55 percent federal tax bite. Hard to believe, but true. The company that did everything in its power to keep from paying corporate income taxes faced the prospect that its founder's estate would pour half its wealth into the sinkhole of the U.S. Treasury. Over the years, Magness declined to consider his own mortality and delayed his estate planning. When Sharon had raised the subject, he said, "Girl, you'll figure it out."

Before he died, paperwork was drawn up to create a foundation with at least $100 million—an idea Sharon said they had previously discussed. But Magness lost consciousness in the hospital before he could sign the papers.

After the funeral, when she went up to her husband's eleventh-floor corner office to clean it out, Malone wandered in and said, "Nobody's watching my back anymore."

The Magness family and Malone faced a dicey situation. Settling estate taxes within nine months of death, as required by law, meant that much of Magness's 32 million shares of TCI stockholdings, mainly in supervoting B shares, would have to be sold. Malone had the first right to buy them, but it was by no means clear what would unfold.

TCI, meanwhile, was in trouble. Malone cut 2,500 jobs, sliced executive salaries 20 percent and put TCI's four jets up for sale. He told Sparkman that he feared losing the company if dramatic action was not taken. The stock ticked up. TCI spun off its ownership in Primestar into a new company, TCI Satellite Entertainment, Inc., (TSAT) with Malone as chairman and major stockholder.

Malone attended the Western Cable Show in Anaheim, California, in December, appearing on a panel with Ted Turner, Craig Mundie of Microsoft, cyberspace guru Howard Rheingold and Kim Polese of Marimba, a software firm. It was four years after Malone had touted the 500-channel future, but his focus this time was on a 50-channel present that could keep TCI paying its bills. He said cable's long era of ever-increasing debt and bottom-line losses was over.

"The story gets a little old," he said. "We've been playing that game in this industry for 30 years." Malone described a new, fiscally conservative TCI, split into programming, international, core cable and telephone companies to clarify its debt structure. "The curse of the cable industry is that every year debt goes up and all the money gets re-invested," he said. "That's a paradigm that has to change.

The industry has to generate earnings at some point if we want to continue to be able to raise capital."

Earnings? Did he say earnings? Some of the attendees couldn't believe their ears. It was as if Fidel Castro had abandoned communism in Cuba—a country that, it turned out, had a gross domestic product in 1995 roughly equal to TCI's debt. Yet, there it was. The chess machine had a new view of the board and even a clever story to illustrate the shift.

> It was this way for the pharaohs in Egypt, you know. The pharaohs were able to build those pyramids because people were willing to set aside some of the grain and not eat it all. The pharaohs understood the idea that you don't take all of the grain and feed it to the workers or replant it as seed. You have to hold some back for new investment on which there is high return. In their case it was immortality. All that Ted and I ask for is a rising stock price.

When Turner spoke, he once again mentioned Adolf Hitler and Rupert Murdoch in the same sentence and opined that Murdoch "doesn't have a friend in the world" and "doesn't treat anybody well, not even in his own company." Turner said he was mystified by the historic lows of cable stocks and thought they would bounce back from the perception that small-dish satellite TV was a cable killer. "We've felt the worst of it, kind of like France did during the opening months of World War I," said Turner. "And then the trench warfare begins."

Malone later added an ominous warning that programmers would have to check price hikes or suffer the consequences. "We want all programmers to consider what they would do without 18 million subscribers," Malone said, referring to TCI's reach through its own cable systems and partnerships. "That should center their thinking."

———

SIX WEEKS LATER, singers Don Henley, John Mellencamp, Jewel and Tony Rich trooped into the atrium at the Denver Center for the

Performing Arts. They took their seats at a long table on a riser, facing a cadre of the press, curious fans and handlers for the Video Hits One cable network. "Things have changed a little bit in the past couple of hours," Henley said into a microphone, sounding as if he were just back from the front to announce a long-awaited treaty.

In the context of home entertainment battles, that's exactly what he was doing. On January 1, 1997, VH1 had been pulled from two million TCI homes. Its sister network, MTV, was off in another 700,000 homes. TCI dropped networks all across the country that day, replacing them with ones that made more money to bolster the company's finances. They reaped a New Year's whirlwind.

"Help, we're trapped by the corporate fascists from the planet TCI," wrote a viewer in Lexington, Kentucky, griping about the loss of Bravo, WGN, VH1 and Comedy Central. There were thousands more angry brushfires across the cabled plain, heating up phone lines, newspaper columns and TV broadcasts.

Vitriol poured off Internet sites like "TCI = Totally Cheap and Indifferent" and "Iowa Held Hostage: We Want Our MTV." It boiled down the angry difference between network television and a networked computer. With a one-way TV tube, viewers silently fumed or griped on the phone. The Internet was a bulletin board for cyberbile. "These SOBs have decided to remove MTV from their lineup in Altamonte Springs," wrote a resident of that Florida town. "Excuse me? MTV? Isn't that one of the channels that MADE CABLE TV IN THE FIRST PLACE?"

In the broadcast, cable and satellite TV industries, viewers were known as "eyeballs" for rating points, advertising dollars and subscription fees. But the eyeballs were getting angry.

In the Colorado mountain town of Salida, TCI took off C-SPAN, the Weather Channel and one of two PBS stations in order to make room for Fox News, MSNBC and the Cartoon Network. So furious were the townsfolk that a Rotary Club member led a petition drive that drew 1,700 signatures. Ed Quillen, a resident of the town and columnist for *The Denver Post*, lambasted Malone on the opinion page of the newspaper. "We are not treated like paying customers;

we are treated like livestock to be delivered to a slaughterhouse," he wrote. "If they really want to show me what I'm looking for, show me chairman John Malone in rags holding a tin cup and a 'Will work for food' sign, standing outside the headquarters, which the sheriff has padlocked in preparation for the auction."

TCI's chief spokesman, Robert Thomson, wrote back to the newspaper, explained some of the network shifts and asked, "Do you really think financial ruin for John Malone and [TCI employees] is the answer?"

Another reader weighed in: "Supply-and-demand laws will eventually lower prices as all the unhappy people cancel, and then you all can sign up again for cheap cable and watch all the garbage you want. Better yet, cancel your subscription and get off the couch."

Malone had inadvertently triggered an economics debate. He was wielding the cable pipe like a truncheon, pushing aside lesser-viewed networks for ones that would pay up-front subscriber fees. Murdoch had paid to get Fox News on, and others were going to do it, too. Payment for access goes on in many walks of life. Cereal makers pay grocers for eye-level shelf space. Computer manufacturers pay resellers to promote their hardware. Donors give money to politicians. The practice goes by many names. Marketing funds. Pay to play. The cable gang called it key money.

Cynics called it a bribe.

During a panel discussion at a trade show in New Orleans, ABC economics reporter Robert Krulwich confronted Discovery Networks president Johnathan Rodgers about Animal Planet's fortuitous arrival into millions of cable homes in 1997.

"How did you get John Malone to give you such a wonderful ride?" Krulwich asked.

"The game is distribution," said Rodgers.

"You bribed him," said Krulwich.

"No, we didn't bribe him," said Rodgers. "We offered him quality programming."

"And then," said Krulwich, cackling, "you laid a little cash in his hand."

Since TCI owned 49 percent of Discovery Networks, the $5-per-subscriber key money paid by Animal Planet was to some degree an internal transfer. The key money from independent programmers like Murdoch was gravy.

The losers, as Sumner Redstone found out, were Viacom channels like MTV and VH1 bumped off to make room. Redstone had not endeared himself to Malone when he filed the antitrust lawsuit against TCI in 1993, the one that accused Malone of "bully-boy tactics and strong-arming of competitors." The one that got the Justice Department poking around in the relationship between General Instrument and cable.

Malone and Redstone appeared to bury the hatchet, or at least leave it on the table between them, when TCI bought Viacom's cable systems in 1995 for $2.3 billion in cities including Seattle, San Francisco and Nashville. As part of the deal, the lawsuit was conveniently withdrawn. That Malone would put money in Redstone's pocket at all surprised some people. Malone was not among them. "I'm a businessman," he said.

Having shed its cable distribution, Viacom was one of the largest programmers not affiliated with either a broadcast network or a cable company. Its leverage was brand names such as MTV, VH1 and Nickelodeon. Redstone thought he had Malone's word to keep Viacom programming flowing through TCI's cable pipes. But with VH1 and MTV banished from a combined 2.7 million TCI homes in early 1997, that was clearly not the case. "I had an absolute, positive commitment from him," Redstone told two Viacom executives during a lunch meeting at the company's Manhattan headquarters. "He promised it would not happen, but it did happen. I do not believe John Malone would cross Viacom. I just don't believe it."

VH1 began a national campaign against TCI, recruiting radio stations and rock stars. Full-page newspaper ads urged fans to flood TCI with phone calls. Redstone himself put in a call—to Malone. On the morning of the VH1 press conference in Denver, EchoStar employees set up a booth outside the event and pressed yellow Dish Network flyers into the hands of everyone who walked by. Inside,

Don Henley and his cohorts prepared to bury TCI as a shameless example of corporate control of the media. This, even though Viacom—the rock stars' benefactor—was no less corporate than TCI and exceeded the cable company in revenues by several billion dollars. Sensing a public relations meltdown in the making, TCI backed down before the press conference and agreed to put VH1 and MTV back where they were taken off.

"Clearly, the voice of the people matters," Henley told the crowd, mission accomplished. "It matters to VH1. It matters to MTV. It matters to TCI." His reverie was interrupted by the trill of a cell phone. "And it matters to whoever that is calling on the telephone. I think this is a testament to the power of music, to the power of democracy," Henley continued, searching for a third power, which was "the power of consumers."

One of the consumers was Mellencamp, looking in dire need of a shave and a cigarette. He said that he was a TCI customer in Indiana and that "VH1 was taken off in Bloomington." A VH1 flack stepped up to a microphone to discuss VH1's ratings, its upset viewers and TCI's callous replacement of the music video channel with networks like Animal Planet. Mellencamp brought her comments, and the press conference, to an end. "At this point, it doesn't matter, because the problem's solved," he said. "Whatever they were replacing it with, who gives a shit?"

But that was the point: A lot of people did. They were either loyal to networks already on the air, or eager to get something on that wasn't, like the Sci-Fi Channel. Angry eyeballs from Alabama to Washington State objected to a slew of network switches, sometimes using the language of junkies. "TCI is like a powerful drug dealer," wrote one viewer to an Internet site. "They've got what you want and they know that you will pay for it. Let's face it. TV is a drug, a necessity for most."

It was video crack, smoked through the eyeballs, with a gangland struggle to control distribution and prices. It was satellite versus cable. And cable versus programmers. Yet, Malone's leverage against programmers and his threats to pull them off the air had the ring of a supermarket railing against rising wholesale prices for cereal, even when it got the biggest discounts and even owned some of the major brands. TCI's Liberty Media held stakes in many popular networks and was worth billions of dollars, with Malone its chairman and a major shareholder.

Liberty's day-to-day chief from 1991 to 1997 was Peter Barton, who ruled with mercenary humor. He once chided a reporter who had written a profile of him, "You didn't go for the jugular." Liberty was TCI's biggest supplier of programming, but also sold to other cable and satellite companies. "We're a little bit like the French," said Barton. "We're willing to sell arms to all people in the war." Liberty owned two-thirds of MacNeil/Lehrer Productions, which produced the famed nightly news hour on PBS. "A welcome infusion of capital into the News Hour," PBS president Ervin Duggan called it. When the Republican-led Congress proposed cutting government support and privatizing PBS—a proposal later abandoned—Barton made overtures to Duggan about Liberty buying PBS outright.

Duggan declined the help and alluded to a cautionary fable: "When the fox said to the gingerbread boy, 'I'll help you get across the river,' what did the fox really want?"

Peter Barton was the fox. He worked in a building separate from TCI's headquarters, though it was still among the corporate mirror boxes of the Denver Tech Center. Married with three children, Barton remained a rambunctious kid himself well into his 40s. At work, he dressed in mountain casual attire: slacks, a pastel blue work shirt and a sleeveless black fleece vest. His wiry black hair was askance. His office looked like a tree house. There were bottles of Blenheim spicy ginger ale, a present from the magician Penn Gillette; boxes of ibuprofen stacked against a wall—a gag gift meant to ease Barton's joint pain from downhill skiing; and a yellow banana costume from the Allen & Co. Sun Valley media mogul conference, signifying Barton's role as second banana to Malone.

On the cluttered desk was a framed picture of Barton with Robert Redford and a book of little-known writings by Benjamin Franklin, titled *Fart Proudly*. At hand was a small paintball gun, from which Barton occasionally shot red and yellow pellets onto a sliding glass door that opened onto a deck. Sitting outside, wearing dark sunglasses and staring off into the Rocky Mountain foothills, he took calls from the likes of Barry Diller and Michael Eisner.

Barton was a made-for-cable mogul. After graduating with an economics degree from Columbia University in 1971, Barton knocked around the West and skied competitively. He returned to the East Coast and landed a job as a top aide to Governor Hugh Carey of New York. He then attended Harvard Business School and, before graduating in 1982, sent letters to several hundred CEOs seeking a job. Malone answered and they hit it off. Barton was one of the few TCI executives—or anyone else, for that matter—who had been to Malone's home. Malone and Barton once took a week-long sailing trip from Maine to Florida. Malone, so guarded and calculating, seemed to open up around Barton. They were cable's Butch Cassidy and the Sundance Kid.

As the programming battles blazed away in February of 1997, Barton was in the thick of it, using the phone as a tool of seduction and pressure. The joke was that Barton's version of digital compression involved thumbscrews. When he hung up the phone, he spun around in the chair and began his discourse to a reporter:

Viewers equal eyeballs. Eyeballs equals revenues. Do you know what Q ratings are? It's sentimental value, brand recognition. The Globetrotters have a higher Q rating worldwide than the Denver Nuggets. Probably also true in Denver. But when you start looking at Q ratings and say, "Well, the brand recognition of a channel like WGN isn't so high." Shame on them. Frankly, shame on me if one of my services ends up being disposable. That's my fucking problem. Not TCI's. My job is to maintain my franchise with consumers so that whenever surveys go out, they say, "I gotta have that one." The untouchables are Fox Sports, CNN, ESPN and MTV to a certain enthusiastic group. You cull from the bottom of the herd and breed up.

So what's the point of key money?

"There was kind of an auction," Barton explained. "But if you have a really lousy service, something that really sucks and you want to pay the cable operator a lot of money, you're not going to get on. Your most important relationship is with your customer."

It was critical, Barton once said, to give the dog the dog food it wants. "When John Malone or anybody else stands up and says, 'Programmers aren't going to be able to raise their prices much anymore' and 'I'm also going to charge key money to get access to the shelf,' there's nothing mysterious about that. He's just looking at it from the distributor's point of view and saying, 'Gee, this shelf is so valuable, I might as well charge rent.'"

Not everyone in the TCI organization was so purely pragmatic. There was Roy Bliss Jr., the son of the Roy Bliss who built the cable system in Worland, Wyoming, in the 1950s. Roy Bliss Jr. worked at the Worland cable system as a kid, burning his fingers changing blown amplifier tubes on the fly. He followed his father into the cable business, worked with Gene Schneider in Wyoming and ended up heading a company in Tulsa, Oklahoma, named United Video Satellite Group, or UVSG. It supplied programming to owners of big satellite dishes, created an on-screen programming guide and distributed the WGN Superstation to markets outside WGN's home base of Chicago.

TCI threatened to pull WGN out of five million homes in early 1997. This seemed counterproductive, since TCI owned 40 percent of UVSG's stock and controlled 85 percent of its voting power. TCI's technology chief, Larry Romrell, sat on the board. Yet, with key money for new channels being offered, TCI's cable division would do better unloading WGN and its steep copyright fees.

Roy Bliss Jr. went on the attack. "This is obviously a decision made by corporate executives out of touch with local markets," said Bliss. "Once TCI's corporate decision makers hear from their local managers and viewers, this decision will likely be reconsidered in many markets."

Though TCI was often a target of criticism and investigation— "the favorite subject of all proctological exams," said Barton—criticism from insiders made public was rare. It violated *omerta*, the code of keeping quiet and rebuffing outsiders. The cable industry was inbred. You never knew who was listening or whom you might be working for—or against—down the road. Taking on Malone publicly was professional suicide. But Bliss was fed up. He said later that he already had one foot out the door. He expressed the feelings of any number of subscribers angry at their cable company for taking them for granted.

"TCI is a big company, with lots of tentacles," said Bliss. "And nobody is quite as cavalier with its customers as TCI." One week after sounding off, Bliss resigned, his departure a symbolic beheading for violating *omerta*.

"He got sacked," said Barton with a shrug. "If you're the kind of a guy that if they hand you the ball, you turn around and run in the opposite direction and put it in the other guy's goal, people aren't going to want you around."

During the conversation in Barton's office, a CNBC interview came on TV, showing a San Francisco–based cable executive named Leo J. Hindery Jr. being interviewed. He had recently been named TCI's president by Malone. Barton watched silently, impassive.

Butch and Sundance weren't going to be together much longer. There was a new gun in town.

"The big story is Leo," Barton said finally. "What's he going to do? How's he going to do it? Why's he doing it? You're going to find that he's an intelligent life form with a good sense of humor. It will be good news."

———

ACROSS TOWN, a TCI subscriber named Kenneth Gibson had never heard of Leo Hindery, Roy Bliss junior or senior, Peter Barton, John Malone, Charlie Ergen, key money or Q ratings. But he did know that WGN was off the air, meaning that he would not be enjoying his springtime ritual of watching Chicago Cubs games.

"Without WGN this spring, I'm going to get mad," said Gibson, a former postman. He lived alone, spending $50 a month on cable, more than for either his phone or utility bills. A bowling bag in the corner collected dust. He flicked the remote control to Channel 51, where WGN and the Chicago Cubs games used to be, but it showed a different type of cub on Animal Planet. He looked out to his tiny patio and wondered about buying a satellite TV dish. "Do you think the cable industry has gotten a little bit too big for its britches?" he asked. "Once they get these franchises, is it a license to steal?"

"THE ONLY WAY you can run a company like ours and do well is be tough," Malone said in January of 1997. "You can't be a social experiment. The message is: I don't give a shit. Toughness is back."

Around this time, Malone and Murdoch met in Denver to discuss their mutual business interests around the world. Also in the meeting were Barton and Chase Carey, the chief operating officer of News Corp. Topics included satellite TV in Latin America, Pat Robertson's Family Channel—which Liberty part-owned and Murdoch had his eye on—and Fox Sports, the multibillion-dollar partnership between Liberty and Fox that was challenging Disney's ESPN.

Murdoch mentioned that he was going to announce a partnership with Ergen to launch a satellite TV service in the United States. Murdoch and Malone had been unable to strike their own deal because of the tangled Primestar ownership.

Murdoch's heads-up to Malone was part of the courtesy of cutthroat: telling your partner about the size and shape of the blade headed his way. Murdoch's alliance with Ergen was coming at the worst possible time. Malone and the cable gang were clawing out of a deep pit, while Murdoch was swinging a sword at their fingers.

"Rupert told us he was going to do the deal with Charlie," Barton recalled. "I remember looking at him and telling him he was nuts. John said, 'Let's move on.' But John was quite pissed about this. It colored the meeting."

ELEVEN

CALLING

DR. KEVORKIAN

CHARLIE ERGEN'S DREAM of rat-zapping the cable gang was coming true. After selling big satellite dishes from the back of a truck trailer in 1980, after betting his company on the nose of a Chinese firecracker in 1995, he was finally playing in the big leagues with Rupert Murdoch—against John Malone.

A few weeks after the Deathstar announcement in Hollywood, Ergen and a contingent from EchoStar headed down to one of his old card-counting haunts, Las Vegas, to attend the spring Satellite Broadcasting and Communications Association trade show. The days of satellite shows in Holiday Inn meeting rooms with butcher paper on card tables were long gone. Glitzy programmers like HBO dominated the convention floor. DirecTV and its equipment vendors sponsored a $100,000 private party with free booze and food on the faux city streets of the New York, New York casino. Even EchoStar, famously tight with a buck, upped the ante of its dealer promotions, offering a $1 million bonus and sports car.

Ergen was surrounded by dealers and well-wishers on the convention floor, many of them still buzzing about the deal with Murdoch and what it all meant. EchoStar displayed a mock-up of its new Sky dish with a dual feedhorn. This would allow the dish to capture several hundred channels at once from the adjoining 110 and 119 satellite slots. During a panel discussion, Ergen stuck out like a casual thumb in the eye—the only one of five executives not dressed in a suit and tie. He wore khakis and his trademark button-down oxford shirt with the Dish Network logo over the left breast.

"Charlie Ergen looked Rupert Murdoch in the eye and said, 'Show me the money! Show me the money!'" said Eddie Fritts, chief lobbyist for the National Association of Broadcasters, one of the panelists. Fritts's membership feared satellite TV almost as much as cable, and for the same reasons. NBC, ABC, CBS—and particularly their local affiliates—worried that their share of eyeballs would continue to erode. Seven out of 10 viewers watched broadcast TV in the early 1980s. By the mid-1990s, it was closer to four in 10. Hundreds of satellite channels would only make it worse. Murdoch's Fox TV was in a different category. The pirate wanted to seize the future with Deathstar.

Life was so much simpler for the Big Three when they had only each other to compete with. They could not reverse viewing trends, but they could use copyright law to block satellite TV from delivering their broadcast channels. The issue was tangled in the courts and Congress. The panelists kicked this topic around, suggesting off-air antennas and bare-bones cable service as a solution for satellite TV's local channel problem.

Ergen had a faraway look in his eyes during this debate. Sky intended to deliver local broadcast signals by satellite into their home markets, beginning with the major cities. The strategy was called "local into local." Ergen claimed to have the satellite capacity, the regulatory authority and the technical ability to do it. When his turn came to speak, the familiar slide flashed on a screen with the word "Cable" circled and a red line slashed through it.

"We just don't want anything to do with it," said Ergen, shooting a dismissive glance to his left at Jim Gray, the chairman of Primestar and former Time Warner Cable executive. "Cut the cable and throw it in the trash."

It was independence day for people who paid to watch TV, according to Ergen. Within a year, Sky intended to match everything cable provided, plus offer broadcast channels and extensive pay-per-view options—all in digital video and audio signals at a price that would undercut the cable gang.

Some, particularly Malone, suggested that Ergen and Murdoch would never get along, that each was too independent for Deathstar to succeed. To quash the doubts, another odd couple made the rounds of the Las Vegas satellite show: Preston Padden, Murdoch's newly named chief of worldwide satellite operations, and Ergen's wife, Candy.

"Charlie has got an incredible entrepreneurial skill set—those are Rupert's words," said Padden, working the media room.

"Everybody in the press keeps saying it isn't going to work," said Candy. "And one of the reasons it can work is because Charlie cares about learning. And he can learn from Rupert." She paused. "As long as Rupert lets him. And so far he is."

There were already cracks in the unified front. Absent from Las Vegas was Carl Vogel, Ergen's second in command. Ergen's decision not to name Vogel president of Sky was the last in a series of rifts between the two men. Vogel quit the company soon after the Sky announcement. He took a job heading a Canadian satellite company, commuting by plane from Denver.

The pressure on Ergen intensified. Making the transition from a growing, but relatively tiny, satellite TV company into a Rupert Murdoch production required an overhaul in operations and vision. EchoStar would continue selling its current satellite TV system for a time. But the marketing would stop, and the Dish Network name would eventually disappear, while EchoStar's manufacturing plants retooled to produce the new Sky dish with dual feedhorn and other components.

"The Dish Network brand becomes a collector's item," said Ergen.

He and Murdoch had already worked the halls in Washington, briefing FCC chairman Reed Hundt and congressional leaders on their deal. They weren't going to make the same mistake Malone had with the 110 slot, snagging it first and worrying about government relations later. Sky faced a host of regulatory hurdles, including foreign ownership, cross-ownership and antitrust questions aimed at Murdoch. DirecTV challenged Sky for controlling two of the three full DBS slots. Broadcasters and the cable gang piled on with their own objections.

Murdoch and Ergen relied on a one-word defense: competition. They promoted it in a full-page ad in *The Washington Post:* "SKY. Finally—the cable competition the Telecom Bill was designed to create." That word, if sincere, could unlock many doors.

The 1996 Telecommunications Act promised that cable, Baby Bell and long-distance phone companies would engage in a market free-for-all, lowering prices and passing on the savings to consumers. No such thing happened or was likely to happen. Instead, many of the local phone companies used the new law to do the dullest, most rational and predictable thing imaginable: They merged back together into what began to look like the old Ma Bell. Bell Atlantic bought Nynex for $26 billion in 1996, and then pursued GTE in a $65 billion deal. SBC Communications bought Pacific Telesis for $17 billion in 1996 and Southern New England Telephone, then pursued Chicago-based Ameritech in a $62 billion deal. BellSouth and US West were the orphans.

Beyond the cost savings and comfort in joining forces, the merging local phone monopolies sought to capture golden traffic: billions of dollars in long-distance phone calls in their combined regions that they could potentially steal away from AT&T, MCI and Sprint under the new law. The Bells were more interested in the $100 billion long-distance market than the $25 billion cable TV market that tempted them earlier in the decade. Malone predicted as much in 1994. "There's nothing the phone companies would like better than for all this info highway stuff to go away," he said. "All they want is to retain their local monopoly and be able to go after the long distance marketplace."

Bingo.

The upshot for consumers of the 1996 Telecommunications Act was higher local phone rates, higher cable rates and few prospects

of effective competition to lower either one. That's why the Sky deal
was likely to find a grateful reception in Washington. It gave political cover to all of those, including consumer advocates, who lobbied for an overhaul of the 1934 communications law in the first
place.

"There's no one who can say this deal is not good for the American public," said Ergen, commingling his commercial interest with
a policy pitch for Sky.

Then a storm cloud floated over from the U.S. Supreme Court.
By a 5–4 vote, the court upheld must-carry rules that required every
local cable system to carry every local broadcaster. "Simply stated,
cable has little interest in assisting, through carriage, a competing
medium," Justice Anthony Kennedy wrote for the majority.

When the case was heard before the court, Acting Solicitor General Walter Dellinger said that without must-carry between 1985
and 1992, broadcast stations were dropped by cable systems about
8,000 times. Justice Stephen G. Breyer noted that only 31 of those
stations went out of business.

Breyer was the swing vote, and yet, he swung for must-carry. The
argument was that it protected viewing options for the one-third of
American homes that did not subscribe to cable. In other words,
smaller broadcasters needed the guaranteed reach of cable, and advertising revenues, to deliver their signals to those without cable.
Basically, a subsidy program.

Even if must-carry had been overturned, cable companies were
not going to drop the local feeds of ABC, CBS, NBC, PBS and Fox.
Viewers wanted them. What might get the ax would be locally-
based foreign language, religious and small public TV channels. Yet,
forcing cable companies to set aside one-third of their channel capacity for local broadcast stations meant that certain cable systems
would not have room to carry the History Channel, C-SPAN or
some other national cable network.

The must-carry subject was so convoluted that while the nine
black-robed sages upheld the law, they tossed the ultimate question
back to Capitol Hill, saying that "competing economic interests ...

in the complex and fast-changing field of television are for Congress to make."

And, since Congress is beholden to the broadcasters, that was that.

Broadcasters were ecstatic over the Supreme Court ruling. The 1,600 network affiliates and independent stations nationwide were guaranteed access through the cable lines. The cable gang had battled must-carry in the courts for 13 years, led by Ted Turner, claiming it infringed on their free speech rights. But they reacted oddly to the loss in court, somewhat like the euphoria when a tooth is yanked but before the Novocain wears off.

They immediately recast the ruling as reason for Murdoch and Ergen's Sky to deliver all 1,600 broadcast stations by satellite into the respective local markets. And, because the enemy of an enemy is a friend, the cable gang and broadcasters found common cause on this vital issue of must-carry for Sky—all in the name of fairness and protecting the public interest, of course.

Padden, who had claimed that Sky would force cable companies to call Dr. Kevorkian, called it a "cynical attempt by cable of thwarting the competition they abhor."

———

LEO J. HINDERY JR. arrived at TCI with his left wrist broken in four places and the ligaments torn from his fingers, following a nasty slip on a heavily polished floor. Hindery lived and worked in the San Francisco area, where he ran InterMedia, a cable company, and lived with his wife and daughter.

Wrapped in a cast, supported by a sling and throbbing painfully, Hindery's busted arm was a good visual metaphor for TCI's self-inflicted damage.

With Bob Magness gone, Malone had stepped up to become chairman of TCI and kept the title of chief executive officer. He stepped down as president and brought on Hindery, whose company was part-owned by TCI. Hindery's appointment came as somewhat of a surprise, since Malone watchers once assumed the heir apparent was Brendan Clouston or Peter Barton or some other lieutenant from within TCI. Malone, perhaps instinctively, chose someone who complemented his weaknesses. Where Malone was aloof, Hindery was engaging. If Malone was a chess machine, Hindery was a cheerleader.

Words spilled from Hindery's lips that TCI subscribers had rarely heard from a senior TCI executive. Words like, "If your perception of TCI or the cable industry was formed because of a rough experience years ago, give me another chance. If it was formed yesterday, I apologize. We'll do a better job. I promise you."

Hindery possessed a weapon absent from the TCI arsenal: humility. He was humbled by the opportunity, awed by the responsibility, eager to serve. He laid it on thick, doing five back-to-back press interviews on one day in mid-February 1997, while his swollen left hand pulsed purple. "There is no other job I would take," he said. "I have an enormous amount of affection for John. If he thinks I can help him, I'd like to try. I didn't want to do anything else with my career. This is very exciting. This is a man I have enormous affection for. It's a great company and a leader of our industry."

Hindery had the fervor and insecurities of a self-made man. He grew up in Washington State, picking crops in the summer as a preteen. He claimed to have become financially independent from his family at age 13, working at a Safeway store while he attended high school. He graduated from Seattle University, a Jesuit school, working his way through college with stints in the merchant marines, as a sheet metal worker at Todd Shipyards and part-time for United Parcel service. He received his MBA from Stanford University in 1971.

Hindery worked in a series of executive positions before joining the Chronicle Publishing Company in San Francisco in 1985. Among its media assets, the company owned cable TV properties, which brought Hindery into contact with Malone. With Malone's backing, Hindery set off on his own in 1988 to found InterMedia, which consisted of cable investment partnerships that grew to serve one million subscribers.

Though not previously a TCI executive, Hindery was part of Malone's inner circle. Of 110 companies incorporated in Colorado with Malone as a director, most of them investment shells, Hindery was a director for 18 of them. He was a tireless worker who slept four or five hours a night, lived on airplanes and had taken up race car driving under the tutelage of champion Richard Petty. Hindery proclaimed equal every constituency that touched the cable industry: subscribers, employees, shareholders, government agencies and the press. He backed women's rise to executive positions in cable and supported AIDS causes. While at TCI, he lobbied against Ultimate Fighting events on pay-per-view and urged the Playboy Channel to buy the Spice Channel for $100 million in an attempt to mitigate cable's drift toward hard-core porn offerings.

At trade events, Hindery spoke at length, thanking everyone in the room as if he were running for office. One trade reporter called him the "round mound of sound."

Hindery said the biggest sin for him was to lose somebody else's money because he knew how hard it was to make. He once recalled his early days starting InterMedia with Malone's backing. "I started dialing for dollars, calling everyone I knew," said Hindery. "I told outright lies. I said John Malone will put money in if you do."

But if he was such a good salesman, why did he need to tell outright lies? That was the queasy question beneath Hindery's hail-fellow, well-met smile. He sometimes treated truth as a variable. He was, at bottom, a politician who happened to be a cable industry executive. Hindery obsessed about information leaking from TCI, though one of the biggest leakers at TCI was Hindery himself, dishing stories and trial balloons that buffed his image. But when faced with probing questions from the press, he sometimes gave misleading answers. And, as he grew more comfortable in the presidency at TCI, Hindery's apparent humility took on the cast of egomania.

"I know people. I know them better than anyone knows them," he told one interviewer. "I don't talk about it much, but I never make mistakes." And this, about his arrival at TCI: Malone "needed me. The industry needed me. It was a unique situation."

On that point, he was right. TCI did need him. He saw clearly where they went wrong and how they had to change. Malone managed TCI like an oil pipeline company, doing big deals and buying wholesale on the supply side. Every other value of the company fell behind that, sometimes far behind.

Hindery took the demand side, viewing cable as an inherently local service that required close attention. "TCI was a gas station company acting like a pipeline company," he said. "Running a pipeline business is a pretty easy business—you just turn on a pump. Running gas stations is really a hard business."

Hindery began hiring back some of the hundreds of field marketing positions cut by Malone. A total of 25 executives, including Brendan Clouston and Robert Thomson, left the company as Hindery brought in his own crew. He restructured TCI's field operations into smaller divisions. Then he pursued swaps, sales and partnerships with other cable companies to create clusters of subscribers. The cable industry was a patchwork of 11,000 separate cable systems. Clustering cut costs within markets. It also helped the transition to a national cable pipeline that could carry two-way phone and Internet traffic, which could interconnect with the existing phone network.

The cable gang had pursued clustering in bits and pieces, but Hindery threw his energy and TCI's weight behind it. TCI's subscriber count began shrinking from 14 million to 10 million through partnerships, while the company offloaded some of its bloated debt, cut capital expenditures and sliced overhead.

"The company needs a huge enema," said Gordon Crawford, the mutual fund manager and consigliere to media moguls, in the spring of 1997. "I think Leo's giving it to them. The biggest thing is the corporate culture there of arrogance and inattention to their customers, and also with the municipalities they operate in and with the federal government. They just got in an enormous amount of hot water."

Hindery cooled things down a bit. But then TCI subscribers faced a 7 percent rate hike in the summer of 1997, which seemed a provocation given the rise of the rat zapper and the pending Sky deal.

"It's a damn dangerous game," said industry analyst Chuck Kersch.

—

AS CABLE INDUSTRY executives headed to New Orleans for the National Cable Television Association convention in mid-March, the mood was bellicose and the air filled with fury about Sky. Murdoch, primarily a programmer, had provoked his existing distributors by seeking to become a distributor himself. The cable chiefs circled the wagons against Sky, proposing to block Fox channels from further expansion on cable.

Smaller cable operators already faced three satellite competitors, including the cable gang's Primestar. Some were petrified of Sky. A cartoon in *Multichannel News*, an industry trade weekly, portrayed a small cable operator pondering two options on a blackboard: "Plan A: Hope Murdoch's plans to air local broadcast signals on 'Sky' aren't realized. Plan B: Build altar; say prayers."

Ted Turner, still battling to keep Murdoch's Fox News off Time Warner Cable systems, continued frothing at the mouth at his enemy's satellite assault. Leo Hindery was sublime. "I have no problem with what Rupert does," said Hindery. "Rupert did exactly what I would have done in a similar circumstance."

Yet, there was a problem, as indicated by a meeting Hindery and Malone had earlier with Murdoch in Denver. The consensus, said Hindery, was "Let's not whale on each other."

Murdoch had good reason to heed that advice.

While chasing Sky, Murdoch spent $2.5 billion for the broadcast TV stations of New World Communications and $1.3 billion for Heritage Media Corp., a direct marketing and coupon business. Plus, he was financing the Fox Sports and Fox News start-ups. Murdoch's appetite for expansion was giving News Corp. investors a stomachache. The stock hit a low of $18.38 within a month of the Sky deal.

Murdoch was courting the same risks that nearly sent News Corp. into the arms of its creditors in 1990. The pirate was sailing close to the edge again, but that's where he thrived. Murdoch could smell the possibilities with Sky. More than one million cable customers a year were choosing the rat zapper, taking billions of dollars in present and future revenue from the cable gang. If DirecTV and EchoStar could do that, what could Sky do?

One thing Sky did was roil the programming partnerships between Malone and Murdoch. Malone's Liberty Media considered selling its 50 percent stake in Fox Sports back to Murdoch in early 1997. Malone was wary of the increasing price of sports rights and preferred to swap Liberty's stake in Fox Sports for a share in the broader asset base of Murdoch's News Corp. It didn't help matters that Murdoch suggested putting some of the Fox Sports programming exclusively on Sky, which Malone would not tolerate.

On April 1, Peter Barton announced his resignation from Liberty. After 15 years as Malone's marksman, he was wealthy, burned out and looking to go solo. His departure came amid Hindery's executive purge. "There's a difference," Barton said later. "They were pushed. I jumped."

Robert "Dob" Bennett, Liberty's chief financial officer, took over as president and chief executive officer. Malone remained chairman. The chess machine had his hands full. While Hindery restructured TCI operations, Malone had a conversation with Murdoch, mogul to mogul.

"I certainly did after the public announcement when what's his name [Preston Padden] said, 'Send for Kevorkian,'" recalled Malone. "I said, 'Jesus Christ, Rupert, we're trying to sell affiliation on FX and Fox News and Fox Sports and having you as a partner under these conditions is not an asset.' I think he understood that."

There were other things Murdoch understood, too, like the road to Pat Robertson's Family Channel. Murdoch wanted to buy it and relaunch it as the Fox Family Channel. Disney/ABC and NBC were also in the hunt. Cable channels with a reach of 67 million homes didn't go on the auction block every day.

Malone's Liberty owned just 16 percent of International Family Entertainment, the parent company of The Family Channel. But the chess machine had first right to buy the controlling shares held by Pat and Tim Robertson. Malone could make or break any deal.

"All roads lead to Denver," Tim Robertson said later.

Also headed to Denver was Eddy Hartenstein, the president of DirecTV, and Hartenstein's boss, C. Michael Armstrong, the chairman of GM Hughes, who had never met Malone. The gist of the meeting was whether it made sense to combine Primestar with DirecTV to compete with Murdoch's and Ergen's Sky. It was an open question.

Malone and Armstrong would meet again in the coming year—for much different reasons.

—

ON APRIL 10, 1997, Murdoch appeared before Senator John McCain's Commerce Committee to make the case for Sky. "The upfront costs for Sky subscribers will include only around $50 to buy the dish itself, plus a $50 refundable deposit per converter box and a reasonable installation charge," said Murdoch. "With that, consumers can receive hundreds of channels of digital pictures with CD quality sound."

Toning down Padden's rhetoric, Murdoch modestly projected that Sky might sign up eight million subscribers over five years, which would "hardly put cable out of business. But eight million customers can provide a profitable business for us, a choice for consumers and an improved cable product for those who might be unwise enough to reject Sky service and stick with cable." Murdoch recalled his history of shaking up the establishment. When Fox TV launched, he said, "The critics scoffed so loudly that I still have a slight ringing in my ear." He touted the creation of the Fox News Channel "despite a firmly entrenched competitor in CNN and in another younger but widely hyped competitor, MSNBC."

What Murdoch wanted of the lawmakers were rules to allow Sky to deliver local broadcast channels without restrictive regulation.

"Sky is willing to risk a $3 billion capital investment to bring consumers a better choice now," he said. "If you give us the legal authority to compete, the rest is up to us."

Murdoch's lobbyists, meanwhile, were trying to get a quick fix of the issue by attaching a local channels rider to a House appropriations bill that funded U.S. troops in Bosnia and provided disaster relief in the Midwest. They did not succeed. Even so, McCain seemed inclined to let Sky deliver local channels. He asked about a published comment from Malone that Sky should be scrutinized because Murdoch had the ability to hold programming hostage.

"If I did all those things," Murdoch replied, "I would be taking a lesson from Dr. Malone."

On that same day, April 10, 1997, Malone, Brian Roberts of Comcast, Joe Collins of Time Warner, Jim Robbins of Cox Cable and other industry officials met at Microsoft's corporate campus in Redmond, Washington. The topic of discussion was Bill Gates's pitch to supply software for a new generation of digital cable boxes that would provide TV sets with many more channels, a computer brain and interactive functions. Gates controlled the software engine, Windows, that drove most of the world's personal computers. He wanted that engine built into a wide range of devices: automobiles, appliances, and most of all, the world's one billion television sets. Just a week before meeting with the cable chiefs, Gates bought WebTV for $425 million. Later, in a TV interview, he recalled:

> We were meeting with the cable industry and they were really depressed at the time. They thought Rupert was going to come in with the satellite system and undercut them. They weren't feeling—the whole interactive TV thing—they had gotten so out in front of themselves on that. And they were just wondering, "What's going to happen to the cable industry?" Their valuations were way down because people were sort of impatient with the promises they'd made. They'd gotten themselves under the gun in a lot of ways. So we said to them, "Come on. Moore's Law—the improvement in chips—is great for you. We can take a lot of these pieces—software, what

we've done on WebTV—put them together and build a set-top box that your customers will find exciting."

And so they thought, "Wow, this is great." And we kept telling how they could get some new revenue from this and they kept saying, "Well, you're Microsoft. You're so clever. Do you really—do you really, really think the cable industry has a great future?"

And we said, "Yes."

The image conjured up of Malone and other cable chiefs mesmerized as Gates waved a magic wand had a fairy tale quality to it. But a fairy tale it became, complete with a buried treasure for the cable gang. And Malone, who once called the satellite TV players the seven dwarfs, was about to hand Murdoch—Snow Black in this version—a poison apple in the form of Primestar.

On the next day, April 11, an article appeared in *USA Today* reporting that Malone and Murdoch had met in New York prior to Murdoch's testimony before Congress. Such meetings were not unusual when it came time to move the chess pieces. Murdoch and Malone once flew in from different parts of the world to Centennial Airport outside Denver, pulled up chairs inside an airplane hangar and got down to business. In 1995, they even allowed a journalist, Ken Auletta of *The New Yorker*, to attend a long afternoon meeting at TCI headquarters. Auletta was in knots, not only from the access he had, but from the pressure building in his bladder, a real back-teeth floater. "For four hours I sat there desperate to pee," he wrote later. "Yet I was convinced that if I left the room they'd never let me back in. So I waited."

What Auletta recorded was two men making decisions like heads of state, scanning the globe for cable, satellite and programming ventures.

"We think Asia will take awhile," said Malone.

"Japan is now," Murdoch said.

"Either you guys get in or ESPN will own sports," Malone warned.

"We got to go in and kick ass," Murdoch agreed.

There were no journalists in the room when Murdoch and Malone met in April of 1997. But afterward, two reporters were told that a meeting took place and what was discussed: Charlie Ergen was about to get his throat cut.

"Instead of teaming with EchoStar, Murdoch's American Sky Broadcasting would join with Primestar, a satellite service controlled by the top five cable companies: TCI, Time Warner, US West's Continental Cablevision, Comcast and Cox," wrote David Lieberman in *USA Today*. The story had a nicely worded denial from Preston Padden, Murdoch's satellite chief: "This is somebody's fantasy."

Bob Scherman of *Satellite Business News* reported a similar story, with additional details: The meeting took place in New York on Tuesday, April 8—two days before Murdoch testified before Congress—and Leo Hindery was also in attendance. "It is unclear how Murdoch hoped to cut a deal with TCI and cut EchoStar out of the picture," Scherman wrote.

Rattled, EchoStar released a statement reaffirming its binding agreement with News Corp. Murdoch called Ergen to deny the *USA Today* report, saying he had not had discussions with members of Primestar in the past year. Remarkably, Murdoch left the message on Ergen's voice mail, a piece of evidence Ergen made sure to keep.

Then, Ergen got a call from Ted Turner, who told Ergen that Time Warner chairman Gerald Levin had been approached by News Corp. officials. "He called to say that News Corp. was throwing us overboard and that they were meeting with Levin to talk about a Primestar/News Corp. deal," Ergen said a year later. "I said I couldn't talk, because I was under a binding agreement with News Corp. But Turner said he didn't want Time Warner to do a deal with News Corp."

At TCI, spokesman Robert Thomson's reaction to the press reports of Murdoch's meeting with Malone was gleeful. "Are you suggesting to me that before Rupert Murdoch presented himself to Capitol Hill as a satellite competitor to cable, he was meeting with cable to cut a deal? We don't have any comment."

PADDEN AND ERGEN, two men who had built their careers bashing cable in words, could not find a way to work together to bash it in deed. They debated the cost of linoleum versus carpeted floors at ASkyB's uplink facility in Gilbert, Arizona, and the transfer of equipment between there and EchoStar's uplink in Cheyenne. At an April 17 meeting at EchoStar headquarters, with Murdoch and other News Corp. officials also in attendance, their feud escalated. Padden walked out of the meeting, never to return.

EchoStar and News Corp. had planned to sign their final contract and file for federal approvals by May 1. Instead, on April 25, Murdoch sent Ergen a letter claiming that a range of issues needed to be resolved. The major one, according to Murdoch, was a requirement that Sky use a set-top box security system from a News Corp.–controlled subsidiary called New Digital Systems. This security system was in use by BSkyB in England and DirecTV in the United States. EchoStar used a different one called Nagra SA.

Ergen claimed that the New Digital Systems technology had been "broken" by hackers and was unreliable. Murdoch insisted that it be used. According to Ergen, the security issue was a ploy by Murdoch, who badly wanted out of the Sky deal and could not do so without breaching their agreement.

The mounting uncertainty had dropped EchoStar's stock from $27 at the time of the merger announcement to about $14. Ergen was shaken. He picked up a copy of *Success Is a Choice: Ten Steps to Overachieving in Business and Life*, written by Rick Pitino, basketball coach of the collegiate national champion Kentucky Wildcats. Ergen, who played basketball every day as a boy in Oak Ridge, Tennessee, read it straight through. He needed to get his head back in the game.

On May 1, Preston Padden resigned from News Corp. after six years. "The EchoStar deal left me without a real job," he said. "I have nothing but respect and affection for Mr. Murdoch, but I am out of here." He was soon hired as a lobbyist for Disney/ABC.

That same evening, Padden's longtime nemesis, John Malone, appeared at a farewell party for Peter Barton at the Oxford Hotel in

downtown Denver. To attend the event, Malone missed a black-tie
Cato Institute dinner in Washington, where he was billed as a co-
host. The Cato gala was attended by 2,000 people, including lumi-
naries such as writer William F. Buckley Jr., Federal Reserve Board
chairman Alan Greenspan, *Forbes* publisher Steve Forbes and actor
Kurt Russell.

Socializing in tuxedos with a large group of people at the Washington Hilton wasn't Malone's idea of a good time. At the Oxford,
Malone could mingle among a small group and celebrate Barton's
career at Liberty. Malone wore a dark suit over a yellow sweater.
Making a rare public appearance with Malone was his wife, Leslie.
At one point, as they stood side by side talking to another couple,
she casually put her right arm across his shoulder. It was easy to
imagine them in the same way 40 years earlier, strolling the Connecticut shore as teenagers.

Also in attendance was Gordon Crawford, who praised Malone
and Barton for the fortunes made by investors on Liberty Media
over the years. Charlie Lyons, whose Ascent Entertainment owned
the Colorado Avalanche and Denver Nuggets, also chatted up Malone. Three months later, Liberty invested $15 million in a new
Denver arena for the sports teams.

Sharon Magness was also there, dressed in white. As the party
wore on and the drinks flowed, John Malone departed. A reporter
approached and asked about the Sky deal.

"It's dead," he said simply.

And TCI? What's the endgame?

Malone considered it for a moment. "The shareholders get rich
and maybe the employees end up owning the company."

Four days later, on May 5, EchoStar demanded that News Corp.
make good on its obligation to loan EchoStar $200 million for on-
going operations. The next day, Chase Carey, the chief operating officer of News Corp., flew out to Denver International Airport to
meet with Ergen. Steve Schaver, EchoStar's chief operating officer,
came along at Ergen's insistence, as a witness.

Carey was a former college rugby player who sported a luxurious handlebar mustache. He reportedly told Murdoch early on that he thought Ergen was a "small-time entrepreneur who would never rise to the level required by the venture." Meeting with Ergen and Schaver at the airport, Carey told Ergen that he would need to re-sign as Sky's chief executive officer for the deal to go forward, to be replaced by a News Corp. designee. No resignation, no deal.

What 10 weeks earlier was portrayed as a partnership of equals between Murdoch and Ergen had evolved into a power play by Murdoch to take over the venture.

Then there would be peace. The kind of peace offered to a lob-ster by slowly heating the pot in which it's submerged from tepid to boiling. Ergen thought he was playing in the big leagues with Mur-doch—against Malone. It turned out that Murdoch and Malone were playing him for a sucker.

Ergen declined Carey's generous offer to step down. Two days later, EchoStar filed a civil claim in federal court, seeking the $200 million payment from News Corp. The next day, News Corp. in-formed EchoStar that it would not loan the money, nor would it pursue regulatory approvals for the merger. Six News Corp. em-ployees at EchoStar headquarters left the building for good that day.

That night, EchoStar officials drove to the home of Steve Ehrlich, the chief deputy clerk for U.S. District Court in Denver. They filed paperwork at his kitchen table, suing News Corp. for $5 billion, al-leging breach of contract. At the federal courthouse on Monday, Ehrlich explained the unusual procedure. "The court clerk is always open," he said, smiling tightly above his bow tie.

The $5 billion sum was based on EchoStar's projected profits from the Sky partnership of $1 billion a year over five years. Wildly optimistic perhaps, but a nice fat number.

Deathstar was dead.

TWELVE

SUMMER OF

LOVE

T HERE ARE MOMENTS of tension that hang for what seems like eternity. A child's first step, knees buckling. A pop fly lost in the sun. A mother's last breath. The crescendo of a symphony. The last card in a poker hand. The sexual act.

Then the tension is resolved and life goes on.

It's like that in business, too. Someone gets the promotion they half expected or the dismissal they half wanted. The shareholders get rich or sue because they didn't. The deal gets done, or it doesn't. Someone gets screwed.

The tension is resolved.

The death of Deathstar was like that. The new life it gave was to the cable gang.

After a flurry of news stories about Charlie Ergen's $5 billion lawsuit—few of them on the front pages where the original deal played—the reaction to the meltdown was fairly muted. From the outside, it looked like another overhyped business deal come and gone, washed away in the ceaseless sea of news.

EchoStar was drowning in it. The company filed a document with the U.S. Securities and Exchange Commission disclosing that it might not have enough money to fund operations past midyear. Ergen received condolence calls from vendors and creditors. He said they told him, "We're sorry you died. When are you going to pay us?"

Ergen and EchoStar's in-house counsel, David Moskowitz, held a press call on May 13, 1997, to discuss the company's first-quarter results and future prospects. A reference to the connivance between Murdoch and Malone came early. News Corp. "sent Dr. Kevorkian to cable," said Ergen. "But cable sent Dr. Kevorkian to make a house call back at News Corp."

Ergen described himself as "naive" and "just a country boy from Tennessee."

A business interference lawsuit against Primestar's cable partners was a possibility. If claimed and proven, it could be a devastating legal device, either at trial or as a bargaining chip for settlement. In 1985, a Texas jury awarded $10.5 billion to Pennzoil after it determined that Texaco wrongfully interfered in a 1984 merger agreement between Pennzoil and Getty Oil Co. Texaco ended up filing for bankruptcy.

As Ergen had shown with Dan Garner, he was not afraid to sue when he felt wronged. The death of Deathstar was the third time Malone had outwitted him. This time was the worst. EchoStar was on shaky ground without a partner and had lost three months getting Rupe-a-duped. EchoStar's business plan called for launching a third satellite in October, which would cost $250 million, plus the ongoing costs of running the business, gaining subscribers and trying to regain some credibility in the marketplace.

Forget about taking on the cable gang, Ergen had to find a way to pay the bills. He needed big bucks fast and that meant going hat in hand to Wall Street, where they didn't want to hear about who Dr. Kevorkian was and whether EchoStar might win a lawsuit in front of a Denver jury packed with TCI cable subscribers two years from now.

Two years was an eternity on the electronic frontier. The technological change, competitive pressures and government policy shifts snapped your head back on a daily basis. By this time, the cable industry had 66 million subscribers; DirecTV/USSB, 2.6 million; Primestar, 1.9 million, with its service increasing from 95 to 160 channels with a new satellite. EchoStar trailed the field at 550,000 subscribers. Murdoch was on ice. AlphaStar, a small-dish satellite TV service just begun in Canada and the United States, was headed for bankruptcy court.

The view from Wall Street is that some stars rise and some stars fall and which one are you, Charlie? So, show us the numbers or Tennessee will be glad to get its country boy back.

"TIME WARNER IS not standing in the way of any agreement," the company's chairman, Gerald Levin, told reporters at the company's annual meeting in May of 1997. "There is no agreement. There's nothing to be stood in the way of."

The agreement that did not exist and that Time Warner was not standing in the way of referred to the possibility of Murdoch dumping his satellite TV assets into the cable gang's Primestar and finding peace in the valley. Ted Turner, as always, opposed helping Murdoch find anything but peace in the grave. He was unwilling for Time Warner to crawl into bed with his nemesis through Primestar, Fox News or any other venue, no matter how crisp the offer sheets. There was a lot of back-and-forth on this throughout May.

In New Hampshire, Senator John McCain of Arizona was stumping early for the state's GOP presidential primary in 2000. Addressing the Merrimack County Republican Party in Concord, McCain said that the result of the Telecom Act of 1996 was, "in case you haven't checked lately, that your cable rates are going up, your phone rates are going up, and they are going to go up even more." C-SPAN covered the event. *Multichannel News* headlined the news item "Cable's Nightmare: McCain vs. Gore."

Meeting on May 19 with Goldman Sachs in New York, TCI president Leo Hindery indicated that "he expected news soon that would calm competitive concerns about satellite versus cable," according to an analyst report sent to investors.

The cable gang was about to cascade miracles.

ON JUNE 9, TCI sold systems totaling 820,000 subscribers in the New York metro area to Cablevision Systems in exchange for one-third stock ownership in the company. The $1.1 billion deal sent Cablevision stock up nearly $10 to $44. TCI's rise was more modest, up $1 to $16. Later in the month, Murdoch and Malone invested $850 million through Fox Sports in Cablevision's Rainbow SportsChannel ventures and Madison Square Garden, a deal they had been trying to complete for two years.

Also on June 9, the big foot fell from Microsoft's corporate campus in Redmond, Washington. The Green Giant was in the valley. Bill Gates announced that Microsoft would spend $1 billion in cash for 11.5 percent of Comcast Corp., the Philadelphia-based cable company. The amount of money was ordinary—$1 billion being table stakes in this realm. But the symbolism was extraordinary. After the botched partnerships between Malone and Gates throughout the mid-1990s—two scorpions in a bottle—the Microsoft alliance with Comcast was the first to signal that the cable pipe would be a major conduit for software reaching homes through TV sets and personal computers. Cable stocks began to rise.

"Some days, you wake up, and there is a god up there, and his name is Bill Gates," said Tom Wolzein, a cable analyst with Sanford C. Bernstein & Co.

Industry trade publications referred to a "Summer of Love."

Gates had met with the cable chiefs in Redmond two months earlier to discuss set-top box software. At dinner, Gates harped on a familiar theme—the slowness of cable plant upgrades to deliver high-speed Internet, and therefore Microsoft, services. Comcast president Brian Roberts, seated next to him, suggested that Gates buy into cable companies as a vote of confidence.

Gates chose Comcast, which lit the fuse of cable stocks dampened by the threat of Deathstar. The world's richest man bestowed legitimacy on the cable gang's works in progress. Digital cable provided more video channels. Cable Internet services were making inroads. A handful of cable systems provided telephone service. The problem would come in deploying digital video, Internet and phone service all at once everywhere, switching all the traffic, pricing all the services and making sure it all worked together. That is, the problem was to remake the tree into a wheel that actually rolled. It would take years for that to happen, but the cable convergence train was at long last leaving the station.

And where was Murdoch?

In June of 1997, he was visited by representatives of Microsoft, Intel and Compaq. They demonstrated a top-of-the-line Compaq computer, chockablock with Intel chips, a 36-inch high-definition TV screen and a range of interactive functions. It sounded like something George Gilder prophesied, that computer technology would swallow TV, which he called a "retarded" medium. "The computer industry is converging with the television industry in the same sense that the automobile converged with the horse," he wrote.

What concerned Murdoch was the computer TV's electronic programming guide (EPG), which could highlight programs of the vendor, in this case Microsoft, much like the Windows home screen did on the personal computer. "I am very worried about what Gates is doing," Murdoch told the *Financial Times* of London later that month. "He is trying to place himself between us and our customers and I don't think there is a legal way to stop him."

Murdoch owned *TV Guide,* the weekly magazine of TV listings. Malone owned the dominant EPG called Prevue, which downloaded daily TV listings to tens of millions of cable homes served by TCI and others. Through digital cable service, Prevue allowed viewers to click through listings of other channels or upcoming programs while they watched a current show. The rat zappers had their own EPGs.

The technology moved subscribers closer to video-on-demand, which is what they said they wanted and were willing to pay for. So, if Murdoch wanted protection from the Green Giant's move into this arena, he was going to need Malone's muscle with Prevue. And how could he count on that if he was teaming up with a guy who advised people to cut the cable and throw it in the trash? The cable

convergence train was leaving the station and Murdoch had to decide if he wanted to be in the engineer's compartment with Malone—or walking the dusty tracks where he had already tossed
Ergen.

"We don't want Gates to get control of our picture," Murdoch
said. "That's the real thing, but it may be possible to work out some
deal."

There were many deals working out.

—

ON JUNE 11, two days after the Gates/Comcast announcement,
satellite TV industry officials gathered at the Hyatt Regency in the
Denver Tech Center for the Global DBS Summit. Ergen sat in a high
wingback chair on stage, discussing the state of the industry with
other executives when a Primestar press conference from New York
came in live via satellite. The lights were dimmed. Onto a big
screen, like the Wizard of Oz, came the face of Primestar chairman
Jim Gray, the same man Ergen taunted at the Las Vegas satellite
show three months earlier.

The Primestar partners had big news. Not only were they going
to roll up their operations into a new public company called
Primestar Inc., they would have a new partner, Rupert Murdoch,
fresh from an attitude adjustment. His DBS assets, including the
110 slot and two satellites under construction, were headed for the
cable gang. Murdoch's investment in Primestar would be passive,
with nonvoting stock ownership valued at about $1.1 billion. Mur-
doch's satellite uplink facility in Gilbert, Arizona, would be sold to
TCI—another basement for Dr. No. The agreement included in-
demnification by Murdoch protecting the cable gang from litigation that Ergen might bring against them. Murdoch was exiting the
DBS business again—his third failure to crack the U.S. market since
1983. And just to nail that down in the future, the agreement called
for News Corp. and its partner, MCI, to refrain from competing in
the U.S. DBS business for 10 years.

"We think this is very good news for consumers. If not us, who? Why should we continue to be denied that slot?" said Gray, referring to 110. "There is no question that this is a DBS company that will compete aggressively with cable."

So, TCI Satellite Entertainment, Time Warner, Newhouse, Cox, Comcast and MediaOne wanted a rat zapper to compete aggressively with cable. They could have achieved that goal by welcoming the original Murdoch/Ergen deal. But it wasn't about competition. It was about who controls the competition.

The chess machine.

"It's a win for the industry to get Rupert Murdoch to think of himself more as a programmer in North America than as a satellite guy," Malone said months later. "That's a win for the cable industry, sure."

The arbiter of who would compete in what ways was Malone, because he had the leverage to arbitrate. Meanwhile, he was free to think of himself as an owner of programming, a cable guy, a satellite TV distributor, a cable Internet provider, a phone service vendor—you name it. The world was his chess board.

The federal government, having failed to come to grips with whether cable and satellite TV ownership should be separate to support competing delivery pipes to the home, had by default left it to Malone to decide how the industry should be ordered. Primestar was the pipe he had just jammed into the gears of Deathstar.

The market would decide what would happen to the 110 slot, auctioneer-in-chief Reed Hundt had said. He just didn't foresee that it would be the Malone market.

At the DBS convention, opinion on Primestar's coup was unanimous.

"They're a big cable operator in the sky," said Stanley E. Hubbard of USSB.

"You could regard it as the surrender of Rupert Murdoch," said Mickey Alpert, an industry analyst who headed Comsat's failed DBS venture in the early 1980s.

"You're talking about the most powerful media moguls in the world in this deal," said Ergen. "I think the dangers are pretty obvious to people on Capitol Hill. I'll be interested to hear Murdoch's new testimony to Congress."

Ergen walked off stage and was surrounded by reporters. He rattled the saber of his lawsuit. "I don't remember any clandestine meetings I had with Primestar," he said facetiously, referring to Murdoch. "In Colorado, it's tough to sell the same piece of real estate twice. Hopefully the court will require them to sell it to us."

There was something different about the way Ergen spoke. His jaw was set differently. He talked slower.

He had braces on his teeth.

Ergen had a top front tooth that was slightly set back. But that wasn't the main reason for the braces. A few weeks earlier, one of his daughters was due to get braces. She was scared. To calm her down, Ergen told her he'd get them, too.

When the appointed day arrived, Ergen tried to back out.

"Daddy, that's just like what Mr. Murdoch did," his daughter said.

―

NEWSPAPERS DATED JUNE 12, 1997, carried the news that News Corp. and the Primestar partners had done a satellite TV deal, settled their differences and were moving on. Analysts applauded Murdoch's capitulation as a smart business move. "EchoStar was a black hole where it would require billions of more dollars to become operational," said Rick Westerman, an analyst at USB Securities. "This allows him to limit his investment."

In many of the stories, another deal was also reported: Murdoch was buying Pat Robertson's Family Channel in a $1.9 billion deal. The common link between the two deals was Murdoch. The less obvious link was Malone. He held the right of first refusal to buy controlling ownership of the channel.

The invisible hand—or foot—of the Malone market was at work.

"What was TCI's role in the Family deal?" Malone was asked three months later by *Advertising Age.*

"We were really neutral in it," he replied. "The only thing we said is that while Rupert was working out his differences with the cable industry, on the satellite side, we wouldn't let any deal happen. Once the satellite deal got worked out, we simply took our foot off it and said, 'It's open to the highest bidder, but it's Pat's call.'"

We wouldn't let any deal happen.

Tim Robertson, The Family Channel president at the time, said in a 1999 interview that he had no idea the channel was a bargaining chip in some greater game between Murdoch and Malone. And he said it wouldn't have mattered. The Robertsons wanted to sell, and Murdoch was eager to buy it as a vehicle for his Fox Family network.

Murdoch summed up his decision to bail out of the Deathstar deal with Ergen, using almost the exact same words as Malone. "We were going to be in opposition to the whole cable industry," said Murdoch. "We really had to choose: are we a software supplier or a distributor? The choice was obviously to say we were going to be a software supplier."

Attitude adjustment.

—

IN THE MIDST OF all this, Malone had another troublesome issue to deal with: the $1 billion Magness estate. For six months, the Magness will had been kept under wraps by a Colorado state court at the request of the estate's executors and TCI.

In the interim, Malone scrambled to shore up his control of TCI's supervoting B stock. In April of 1997, TCI acquired the Kearns-Tribune Corp., publisher of *The Salt Lake City Tribune,* by issuing $627 million in new TCI stock. It wasn't a diversification strategy. Malone wanted to get his hands on nine million shares of supervoting TCI stock owned by Kearns-Tribune. The company was an early investor with Bob Magness in TCI. Kearns-Tribune chairman J. W. "Jack" Gallivan, who was in his 80s, sat on the TCI board of directors.

One consequence of the deal was that 24 longtime newsroom employees of the Salt Lake paper received a windfall in their company stock plan, up to $200,000 extra apiece. Malone was willing to spend what it took because of the uncertainty surrounding the estate. The 55 percent estate tax bill was due in mid-August of 1997, nine months after Magness's death. That meant that much of the stockholdings would have to be sold. Malone had the first right to buy them, but he devised a clever solution that didn't require that he personally put up any money. He worked closely with the estate's two executors: Donne Fisher, a board member and former TCI executive, and Dan Ritchie, the chancellor of the University of Denver. Both Fisher and Ritchie were longtime friends of Magness.

The core of the deal involved 30.5 million supervoting B shares held by the estate, representing about 20 percent voting control of TCI. Those shares would be placed in the TCI treasury. Then, an equivalent amount of common A shares would be issued by TCI to two Wall Street brokerages, Merrill Lynch & Co. and Lehman Bros. At a share price of $16.52, the brokerages would loan $529 million to TCI, which would then pay the estate. TCI retained the right to buy back the common A shares from the brokerages at the sales price, plus interest.

Meanwhile, Malone would retain his right of first refusal to buy the B shares, tucked safely in the TCI treasury.

It was the chess machine at his finest. The deal was all set—except the estate's main beneficiaries didn't like the smell of it. Kim and Gary Magness wondered why the estate couldn't itself borrow money against the value of its TCI shares and hold on to them. Or, have the executors shop the stock around and see what price it might bring. After all, voting share control could be worth a big premium. It always was to Malone.

The discussions got hot, with deep undercurrents. Kim and Gary Magness were the blood sons of Bob Magness. Kim Magness was also a TCI board member. Malone had built TCI with Kim and Gary's father, but it seemed to the sons that he wanted it all for himself. The accusations of self-dealing that had dogged Malone for years from outside TCI were now coming from within. The sons accused Fisher of a conflict of interest because of his ties to Malone, his board seat and a $450,000 consulting contract with TCI. They claimed Fisher wasn't looking out for their best interests because he was "afraid of pissing Malone off," according to court documents.

Ritchie said he and Fisher were skittish because they were personally liable for the estate's disposition and tax payments at a time when TCI's stock price was variable.

The view from within TCI was that the Magness sons had done little, if anything, to create the wealth they claimed for themselves and should be thankful for what they got. Indeed, they were due a separate $200 million payout from the estate of their mother, Betsy Magness, which flowed through their father's estate to them.

On the morning of June 16, 1997, as officials prepared to close the stock-swap transaction at TCI headquarters, Kim Magness and his attorney were on the way. They talked to Fisher first, who told them, "Don't stop for lunch because this deal will be closed."

And it was.

But at 6:40 that evening, a two-page letter from Lazard Frères & Co. in New York skittered out of the fax machine at the offices of Holme, Roberts & Owen in Denver, the attorneys for the estate executors. The letter from the investment bank was addressed to Fisher and Ritchie and stated that an unnamed "Purchaser" represented by Lazard was "willing to pay in excess of $20 cash per share for all of the Series B TCI Group Common Stock owned by the Estate and a comparable premium to the current public market price for all of the shares of the other aforementioned classes of stock owned by the Estate," meaning the estate's shares of Liberty Media.

"It is our understanding that the heirs are interested in maximizing the value of the Estate," the letter continued. "As a courtesy to Kim and Gary Magness, we informed them immediately prior to transmitting this letter of our intention to do so and have forwarded copies of this letter to them and their advisors."

Who the hell was this, coming in at the last minute, trying to put the chess machine in check?

It was Comcast, playing cutthroat.

—

FOR 25 YEARS, Comcast was a relatively quiet player in the cable industry. After rising to chief executive officer of a men's accessories company, Ralph Joel Roberts left to start a venture capital firm in Philadelphia. In 1963, he bought his first cable TV system in Tupelo, Mississippi. With him was Julian Brodsky, Roberts's accountant.

"It was a very strange time for a Jewish guy from the North to be wandering around Mississippi at the height of the civil rights movement," Brodsky told the *Los Angeles Times*. "I look like an agitator. I talk like an agitator. Every cop thought I was an agitator."

Comcast bought franchises in western Pennsylvania in 1971 and by the beginning of the 1980s had about 200,000 cable subscribers. Brian, the second youngest of five children by the boss, had a summer job climbing poles and stringing cable, though he had trouble carrying the ladder. "I was too weak," he recalled. "I was too skinny."

He got stronger, grew to six-feet-two, and became an All-American squash player. He showed an early interest in investing and graduated from the University of Pennsylvania's Wharton School in 1981. He worked his way up at Comcast, including a stint as assistant general manager of the Flint, Michigan, franchise. It was there he learned a secret of the cable trade: Revenues increase in a downturn. Auto plant layoffs crippled the Flint economy, but subscriptions rose because cable is relatively cheap entertainment compared to movie tickets and live sporting events.

Comcast and TCI joined in a $2.8 billion bid for Storer Cable in 1988. Comcast was also involved in the bailout of Turner Broadcasting, but to a lesser degree than other members of the cable gang. Comcast moved more cautiously into programming and appeared to be less of a risk-taker than TCI.

Brian Roberts became president of Comcast in 1990 at age 30 and began reshaping the company with guidance from his father, Ralph, Comcast's chairman, and Brodsky, the vice chairman. Roberts was nearly 20 years younger than the chiefs of the other cable companies. He gravitated to Malone, picking his brain to learn the killer moves. During a flight to Japan by cable officials in the early 1990s, Roberts snagged the seat next to Malone and they talked the entire trip. Some TCI executives under Malone referred to Roberts as a tagalong, or worse.

But they underestimated him and the wise eyes behind him.

In 1992, Roberts landed Barry Diller to head up Quality Value Convenience (QVC), the cable shopping channel. Ralph Roberts had backed QVC founder, Joseph M. Segel, in the venture and bought stock when it went public in the mid-1980s.

Knocked by critics as the schlocky intersection of TV and retailing, QVC and its kind were actually innovators. Cable catalog services were one of the few interactive TV technologies besides addressable cable converters for pay-per-view movies and electronic programming guides that actually succeeded on a mass scale in the 1990s. QVC's $1 billion annual revenue stream was not under federal rate controls and the channel was popular among cable systems because they got a 5 percent cut of the merchandise sales in their territories. QVC represented dollar democracy at its finest. Its revenues were its ratings.

Having left Murdoch and Fox behind, Diller was eager to get in on the next big thing. He talked it over with the Robertses and Malone, each of whom owned about 17 percent of QVC. Diller was sold on the idea while watching his friend, Diane Von Furstenberg, sell $1.2 million in clothing within two hours when she appeared on QVC in 1992.

Diller bought a stake in the company and set up shop three days a week as QVC's chief executive at company headquarters in West Chester, Pennsylvania. Diller had much more on his mind than moving merchandise. He intended to use QVC as a platform for his own media fiefdom. That led to some tense moments between the Robertses and Malone.

The first round came when Diller and Comcast used QVC to bid for Paramount in 1994, the movie studio that Diller ran for 10 years. TCI bailed out of the bid because it was negotiating the Bell Atlantic merger. After a six-month battle, QVC lost to Viacom, leading Diller to offer the succinct concession, "They won. We lost. Next."

Next was CBS, which offered to buy QVC and make Diller head of the merged company. With federal rules prohibiting cross-ownership between cable and broadcasting companies, Comcast feared a loss of influence and hijacked the deal, making a $2.2 billion counterbid for all of QVC. This trumped Diller and Malone, who wanted the original deal and were unhappy with Comcast's maneuver. Having frozen out CBS, which declined to get in a bidding war, Comcast then went back to Malone and Peter Barton to get TCI's Liberty Media involved in a joint buyout of QVC. Diller went on to head Home Shopping Network with Malone's backing.

Comcast's young president deferred to Malone as "captain of the cable team." But it became increasingly clear that Brian Leon Roberts had his own game plan.

———

THE FAILURE OF Bob Magness to protect his wealth from the taxman became a cautionary tale in the estate planning business. It was relatively easy to set up tax shelters before death, including charitable trusts that protected the core wealth for children and grandchildren. It got people thinking, particularly if they had any money to think about. The Robertses did. In the first half of 1997, Ralph Roberts began transferring control of Sural Corp.—the entity holding the family's 79 percent voting control of Comcast—to Brian. The move would lessen the estate tax impact when the elder Roberts died. He was 77 at the time, five years older than the age at which Magness died. Ralph Roberts also asked the FCC to transfer

60 cable antenna relay licenses to his son, avoiding the need to do so after death.

Having gotten their own house in order, the Robertses turned their attention to TCI's.

With the June 9, 1997, announcement of Microsoft's $1 billion investment in Comcast, the perceived balance of power in the cable industry shifted toward Comcast and away from TCI. The perception was that Comcast and Microsoft were driving the cable convergence train and that Malone was old school, a cable guy who had failed to sell out to a phone company.

Comcast saw a chance to press its advantage quickly. Hence the letter of June 16 from Steven Rattner of Lazard Frères to the Magness estate executors, offering $20 a share for the estate's supervoting TCI stock. It was a ballsy move that threw the Magness estate into further turmoil and opened the door to a potential takeover of TCI.

Malone saw the Green Giant's grasping hand in all of it, that Gates was using his new partner, Comcast, as a front to try to make its move on TCI at a vulnerable time. "What angered me was the backdoor nature of the attempt," Malone said later. "If Bill's interested in our company, he should come in the front door."

Peter Barton, who had already left Liberty Media by that time, did a slow boil at Comcast's move on TCI. "It's like trying to steal your children," he said.

On June 17, the day after the Lazard letter quietly appeared, TCI announced its deal to control the Magness estate's supervoting B shares, with a payment to the estate of $529 million. There was no word publicly of the objections by the Magness sons, nor of the attempted coup by Comcast and Microsoft.

—

A WEEK LATER, Sharon Magness filed a claim for up to half the estate's wealth. Active in charities, she said she wanted to fund the foundation that she and Bob Magness had discussed before his death. His will left her $35 million, plus property, including their home, artwork and the Arabian horse ranches. Under Colorado law, however, a surviving spouse could claim up to half the decedent's estate, tax-free. This was a potential postmortem solution to the 55 percent tax bite due to the IRS. Had she and Bob Magness's two sons been on the same page, they might have pursued this plan together and divvied up the proceeds. But there was too much animosity between them. Within a month of her claim, the Magness sons filed to block her claim, citing her 1989 prenuptial agreement with their father that limited her payout. The sons later filed to remove Fisher and Ritchie as executors of the estate and to rescind the transfer of the estate's supervoting stock to TCI.

It was a real horse opera.

—

LEO HINDERY WAS high as a kite in the summer of love, circa 1997. His left arm was healed and so was TCI. The company planned to offer digital cable service to the majority of its subscribers by year-end, which would help keep the rat zappers at bay. The cable clustering strategy was working, and a gelded Rupert Murdoch was on board Primestar. Hindery himself got a $36 million payday as TCI bought out his ownership stake in InterMedia.

Hindery was Sergeant Pepper teaching the band to play. His operating philosophy was simple: "Don't out-local the local." He said he got it from Al Neuharth, founder of *USA Today*. Hindery himself had worked as chief planning officer for the publisher of the *San Francisco Chronicle*. Hindery liked to trumpet TCI's comeback in the press to reach his cable customers, employees and the business community at large. No city was more important than Denver, the company's hometown and site of its biggest cable system, with more than 400,000 subscribers in the metro area, many of whom also subscribe to *The Denver Post*.

On July 21, 1997, the newspaper published a story headlined "Hindery and TCI's New Spirit: Cable Giant's President Ponders the Road Ahead." It recounted Hindery's diligent work habits and his character, which he attributed to "the Jesuits and the integrity, discipline and intellectual curiosity they instilled." It described his eleventh-floor corner office—Bob Magness's old office—and some objects in the room: a sign that read "Factotum," Latin for handyman, and a Dr. Seuss book titled *Oh, the Places You'll Go!*

A more accurate title would have been *Oh, the Fibs You'll Tell!*

There were 13 pairs of questions and answers in the story, including this series:

> POST: How important was getting Murdoch into the Primestar alliance and getting Charlie Ergen's proposed satellite-TV partnership with Murdoch off the table?
>
> HINDERY: That had nothing to do with TCI. That was TSAT [a separately traded TCI spin-off], which is a Primestar partner. Charlie and Rupert had an arrangement that didn't materialize. Rupert came to the partners with a transaction that he must have thought was better than the alternative. I don't think it was an earth-shattering transaction ... Charlie, Primestar and DirecTV are all [satellite] providers. Whether Rupert worked with Charlie or Rupert worked with Primestar, there are still three very vibrant and successful providers in the marketplace.
>
> POST: Do you expect Primestar to compete with TCI? It's owned by the cable companies.
>
> HINDERY: Kick my ass any time it wants to. It's trying hard. You think DirecTV is chopped liver? The answer is that the [satellite] business is a very aggressive competitor to the cable business. Don't tell me that Primestar is not a compelling competitor. And TCI has no ownership stake in TSAT and none in Primestar.
>
> POST: Except that TCI and TSAT share the same chairman: John Malone.
>
> HINDERY: John will have no involvement with Primestar going forward. Look, if a customer decides to leave the cable industry and go to [satellite], I lose that customer irrevocably, in my opinion. Charlie Ergen is a formidable competitor. So is General Motors [which owns DirecTV]. I don't feel particularly sorry for either one of them.

His responses were bold and resolute: "That had nothing to do with TCI; Kick my ass any time it wants to; John will have no involvement with Primestar going forward."

But the death of Deathstar had everything to do with TCI and Hindery, as a U.S. Justice Department investigation would show. Meanwhile, TCI sold Primestar in its own territories—it was kissing, not kicking, Hindery's ass. And not even Malone, who was TSAT's biggest individual shareholder and mastermind of Primestar strategy, would vouch for the last answer.

"I don't know why he said that," Malone said later. "I'm a big shareholder."

—

ON JULY 23, Time Warner and News Corp. announced a settlement of their litigation. The Fox News Channel would be carried on Time Warner Cable systems, gaining Murdoch millions of potential viewers in a cable news market already crowded with Turner's CNN and Gates's MSNBC.

"We are happy to have this dispute behind us," Murdoch said.

Ted Turner, speaking later in the summer about his plans to convince cable operators of a new payment scheme for his Atlanta Superstation, said, "What Martin Luther King is to brotherhood, I am to the cable business."

—

SUMMER LINGERED INTO early October. *Forbes* magazine published its 1997 list of the 400 richest Americans. Editor James W. Michaels wrote, "Why don't The Forbes Four Hundred give away most of their money? Are they just plain greedy? (We hear that all the time.) People who ask that question fail to grasp the essence of capitalism: For capitalists, money is a tool. Power springs from it, as from the muzzles of comrade Mao's guns under communism. Your capitalist sees money as an instrument for getting things done."

William Henry Gates III was number one at $40 billion.

Paul Allen was number three at $17 billion: "Teamed with Gates to write software for microcomputer; led to creation of Microsoft. Left 1983 when diagnosed with Hodgkin's disease; recovered. Still owns about 8 percent of company. Also owns Portland [basketball] Trailblazers, Boeing 757, 150-foot yacht, impressionist art collection. Avid scuba diver, musician; jams with buddy Dave Stewart, funding Jimi Hendrix museum."

Ranking 19 and 20 were Barbara Cox Anthony and Anne Cox Chambers, whose Atlanta-based media empire included Cox Cable, $5 billion apiece. Rupert Murdoch came in at $3.9 billion. Ted Turner, $3.5 billion. Sumner Redstone, $3.1 billion. Amos Barr Hostetter Jr., founder of Continental Cablevision in Boston, who sold his company for $11 billion the year before to the US West Media Group, $1.3 billion.

Alan Gerry, who sold his cable company for $2.8 billion the year before to Time Warner, $970 million. He was pictured on the site of the original Woodstock festival in New York State. He bought it for $1 million with plans to develop an outdoor arts and entertainment venue. A photograph of Gerry at the site is captioned, "Once Max Yasgur's farm, now a cable guy's fantasy."

The Magness estate ($960 million in 1996). "Future of estate uncertain, wife and kids contesting."

Frank Batten Sr., who sold Tele-Cable to TCI in 1994 for $1 billion in stock, $915 million.

Stanley Stub Hubbard, $890 million. "Pioneer of satellite television took his United States Satellite Broadcasting company public last year, saw personal holdings rise to nearly $2 billion. Fell back to earth when competitor EchoStar undercut industry prices."

Michael Eisner, $760 million.

John C. Malone, $750 million. His photograph caption reads, "Master of the sleight of hand," a reference to his deal to control the Magness estate shares. "TCI stock up 20 percent since Bill Gates completed $1 billion investment in Comcast."

Lawrence Flinn Jr., retired chairman of United Video Satellite Group, a company controlled by TCI, $710 million.

Harold Fitzgerald Lenfest, head of cable company half-owned by TCI, $700 million.

Michael Milken, $700 million. His brother, Lowell Jay Milken, $500 million.

Charles Dolan of Cablevision, $610 million.

New to the list: Charles Ergen, $500 million.

Ralph J. Roberts of Comcast, $465 million.

On the last page of the magazine were "Thoughts on the Business of Life," including one from George W. Spayth: "Enemies are to be selected with care. Once their usefulness as enemies has worn off, they can be welcomed back as friends."

—

ALSO THAT OCTOBER, Vice President Al Gore addressed the first meeting of a blue-ribbon panel that would make recommendations to the president on the public-interest obligations of broadcasters for the free digital TV spectrum they received. Gore lobbied for free TV time for political candidates. "As this committee deliberates and defines the public interest in the digital age, I urge you to pay special attention to the need for TV time to be set aside free for the survival of our democracy," he said.

At the FCC, Reed Hundt prepared to depart the agency after four years, having previously announced his decision to leave one year short of his appointed five-year term. Commissioners James Quello and Rachelle Chong were also on the way out, to be replaced by a new set of political appointees.

Before Quello left the agency, he recalled Hundt telling him, "I've been a very successful litigation attorney. You are my most formidable opponent."

Replied Quello, "Now what the hell do you want? You are a very gifted manipulator."

Questions persisted about the FCC's role.

A book by Peter Huber was published, titled *Law and Disorder in Cyberspace: Abolish the FCC and Let Common Law Rule the Telecosm.* Huber, an attorney and senior fellow at the Manhattan Institute for Policy Research, reviewed the history of the agency, various court rulings in communications law and the behavior of businesses under regulation. The book concluded that the FCC "should have been extinguished years ago" as it has "protected monopolies, obstructed efficient use of the airwaves, corrupted common carriage, mispriced services, curtailed free speech, weakened copyright and undermined privacy."

—

ON OCTOBER 5, 1997, Charlie Ergen walked toward the white sands of Cocoa Beach, Florida, shortly after dawn, with a Bowie knife in his hand. An air force tradition holds that a sacrificial weapon be buried on the day of a rocket launch to bring good luck. A small group, including Ergen's wife, Candy, and EchoStar officials, stood in a semicircle for the ceremony as Ergen dug a hole in the sand 15 feet up from the Atlantic Ocean's soft roar.

A geezer with a metal detector and earphones came toward the group, scanning the sand. "Probably sent by Malone," muttered Ergen, who jammed the knife to its hilt in the hole and pushed the sand on top.

Ergen kept EchoStar going by digging a deeper hole with debt. His financiers at Donaldson, Lufkin & Jenrette sold another $375 million in junk bonds, plus $200 million in preferred stock convertible to debt. That supplied money for ongoing operations and paid for the $250 million launch of *EchoStar III*, the satellite atop a Lockheed Martin rocket awaiting launch at Cape Canaveral.

"We were given up for roadkill in June," said Ergen. "But you can't back down when the cable bully starts demanding your lunch money. We've got the public's support, to the extent they're educated about it. We're fighting for the hearts and minds of consumers."

EchoStar III seemed a particularly quixotic quest. Ergen planned to use it to deliver local broadcast channels into major markets by satellite, an issue that seemed hopelessly tangled in the courts and in Washington. But Ergen was committed to going head-to-head with cable—"old, analog, rotting miles of cable," he said as if tasting something unpleasant. "Once you've experienced digital satellite, you're not rushing out to buy cable."

Ergen was gambling again, with or without Murdoch, and regardless of whether anyone else thought his strategy was worth a damn. His arrival on the *Forbes* richest list was impressive, but it was a paper fortune backed by corporate debt. In August, the company was notified that it could be delisted from the Nasdaq stock market because EchoStar's net assets were below the $2 million minimum. They straightened out the problem.

At a safe distance from the launch site, late in the afternoon, several hundred EchoStar employees, vendors and their families sat on portable aluminum stands as at a softball game. Ergen's mother, Viola, and several of his siblings were there. It was 40 years to the week that Charlie Ergen and his father watched *Sputnik I* orbit the Earth. Now Charlie had his own satellites to play with. *EchoStar III* rose slowly off the launchpad, then picked up speed, moving out of sight on an arc across the Atlantic Ocean toward Africa on its way into orbit.

"Beats laying cable," said Judianne Atencio, EchoStar's spokeswoman, walking across the tarmac as the EchoStar partisans cheered the launch and hugged each other.

The entourage returned by buses to the two-story Hyatt Hotel complex near Disney World and began partying by the pool. As someone new walked up, or turned to the poolside bar to get a drink, Ergen and his henchmen snuck up, grabbed the victim by the waist and launched backward into the pool, the shriek of abduction swallowed in the silence of underwater. Ergen swam to the edge, hoisted himself up and scanned the crowd for another victim.

Had Murdoch or Malone been there, they too would have been submerged—and perhaps held underwater for longer than is healthy.

The pool party broken up by hotel security, Ergen and a small group retired to his nearby hotel suite. As a poker game ensued around 2 A.M., Ergen's wife and several other women put the last of the Ergen children to bed. Ergen wore shorts, a short-sleeved polo shirt and white socks pulled up nerdishly high on his calves. The room was filled with nervous exhaustion, the aftermath of a 20-hour day that began at dawn at Cocoa Beach.

Sitting around a low coffee table were Ergen; EchoStar counsel David Moskowitz; T. Wade Welch, the attorney handling the Murdoch litigation; and Rick Westerman, the USB Securities analyst who had quit his job in New York, was living in Florida and would soon join EchoStar as its treasurer.

The men, with unlit cigars hanging from corners of their mouths, took turns dealing the cards. Red and blue poker chips were pushed into the center of the table as the betting dictated. During a showdown in seven-card stud, with all but one of the other players having folded, Ergen leaned back to look at his cards, then leaned forward to bet.

He wasn't showing much, a pair of threes and two mismatched cards compared to his opponent's two pairs: kings and eights. Both players had two hole cards unknown to the other. They could choose to get the seventh card dealt facedown for $1. The betting escalated, with perhaps $30 in the pot. Then the last cards were dealt. The opponent flipped his cards, displaying three kings and two eights—a full house—with two garbage cards. Ergen began peeling his hole cards one at a time, prolonging a sure loss. But the last two cards he turned were threes, including the card he bought facedown.

Ergen had a natural four-of-a-kind, beating the full house. A cool hand fluke. But it won and Ergen swept his right arm forward to collect his chips as if he had known it all along.

THIRTEEN

ENDGAME

W HAT RESULTED FROM the high stakes and killer moves
on the electronic frontier? What difference did it make
what a bunch of millionaires and billionaires did with their money
or their technologies—or to each other, for that matter?

Not much, except for the bill presented for the spectacle. In September of 1997, the Consumers Union, publisher of *Consumer
Reports* magazine, called for an emergency cable rate freeze and a
Federal Communications Commission review of the intertwined
ownership of cable, programming and satellite TV assets. It included a greatest hits sampler from the summer of love.

> After Rupert Murdoch abandoned efforts to compete with cable,
> cable industry leaders helped Murdoch win the bidding to pur-
> chase The Family Channel, get carriage of his news channel on
> Time Warner's cable systems and join the cartel of cable owners
> who run the Primestar satellite venture. Now, in conjunction with
> TCI, Murdoch is expanding his sports programming empire by
> purchasing a large share of Cablevision Systems Rainbow Media
> Holdings, combining 18 regional cable sports channels, the Fox
> national TV network, Madison Square Garden, the New York
> Knicks, the New York Rangers and Los Angeles Dodgers in one
> ownership circle.

Feeding the interlocked beast were cable rates rising 8 percent a
year—more than three times the general inflation rate—since

enactment of the Telecommunications Act in February of 1996, primarily because there was no effective competition to keep rates in check. The cable industry claimed that higher programming costs drove the rate increases and that subscribers got good value for the money. Yet, in places where competition did exist—such as the Chicago suburbs where Ameritech, the Baby Bell phone company, wired homes for video service—rates were anywhere from 5 to 20 percent lower.

The force behind the Consumers Union report was Gene Kimmelman, another member of the Tennessee tornado that bedeviled the cable gang. Kimmelman, in fact, grew up in Oak Ridge, the same town as Charlie Ergen. He attended the same high school as Ergen, but was a few years younger and claimed not to have many dealings with him in Washington, where Ergen had become a gadfly. It wouldn't have mattered if they were in constant contact, singing the Oak Ridge fight song, since their professional interests were already aligned against cable.

Kimmelman had graduated from Brown University and was a staff attorney for Ralph Nader's Public Citizen Group. He was the Consumer Federation of America's legislative director for a decade, then spent two years as staff director for the antitrust subcommittee of the Senate Judiciary Committee. He became co-director of the Consumers Union in 1995 and was a familiar face at congressional hearings on telecom legislation.

Kimmelman knew the ins and outs of Capitol Hill, but he was tilting at windmills if he thought the FCC or Congress was going to enact a cable rate freeze. The government had its chance with the 1992 Cable Act. Had there been the foresight and political will to separate cable, programming and direct-to-home satellite TV ownership at that time, there would be two distinct pipes into the home, with independently owned sources of programming—a theoretically competitive market. One difficulty then would be determining how much ownership concentration to allow in programming. In an interlocked business environment, with cross-investments and cross-purposes, there were no easy answers and no clean cuts.

To even head down that road would have required predicting the rise of the rat zapper as a competitor to cable, ending broadcast must-carry and fighting a likely court battle with the cable gang over programming ownership—all of which assumed that such a sweeping menu of intervention would even be supported by Congress, the president and the Supreme Court, where it undoubtedly would have landed.

What Congress did instead in the 1996 Telecommunications Act was to lift barriers between local phone companies, long-distance carriers and cable companies to let them compete. All the major industry players lobbied for the law—and then sent their lawyers into court and Congress and the FCC afterward to lobby against the parts they didn't like. The practical effect of the law was to spur more mergers, with no rate relief for consumers.

The truth is that it's not always economically rational to compete, though posing as a competitor is a great strategy. László Mérō, a Hungarian mathematician and game theorist, wrote about such behavior in the wild.

> Certain animals compete over mates or territory not by fighting, but by standing face to face in a threatening posture, ready to fight, and remaining that way for a long time. Finally, one of them backs down, and the other wins the valued commodity. This method of solving conflicts is common among animals living in strict hierarchical societies, but it also occurs in animals that do not live in groups, rarely meet one another, and have little memory of the outcome of previous fights. This form of combat is particularly common in well-protected animals for whom injury is very unlikely. For them fighting does not make much sense, because the outcome would depend greatly on chance. Furthermore, an injury could be fatal. For similar reasons, animals with strong offensive weapons may also settle their disputes by some kind of posing.
>
> Animals that do not live in groups pay for such posing fights in the currency of time. No matter how valuable the commodity in question, no animal can afford to spend too much time posing—they have other vital things to do.

IN LATE 1997 AND EARLY 1998, John Malone struck a mighty pose. He appeared to be standing in Denver holding up a large neon sign that blinked west toward Silicon Valley and the Redmond lair of the Green Giant, and that blinked east toward AT&T headquarters in Basking Ridge, New Jersey. The sign read, "Let's Make a Deal," just as it had four years earlier before the doomed merger with Bell Atlantic Corp.

In high spirits, Malone attended industry events, chatted up reporters and appeared on the cover of *Fortune* magazine. The dark days of 1996 were long gone. From a low of $11 then, TCI's stock was at $29 and rising. Malone, Leo Hindery and Colorado governor Roy Romer attended an indoor groundbreaking in late 1997 at a business park in Douglas County, one of the booming suburban flatlands south of Denver. TCI was going to build a new $50 million headquarters, shaped like a Pentagon with rounded corners. "None of us had the sense of circularity except John," said Hindery.

Malone himself was circling around the magic box, the holy grail of interactive television. The magic box would be the computer atop, and someday within, the TV set, making it a two-way eye: into the living room and out to the world. Magic box experiments ranged from Qube in Columbus, Ohio, in the 1970s to Time Warner's Full Service Network in Orlando, Florida, in the early 1990s.

The premise was always the same, that television could be transformed from a passive viewing device into a tool for home shopping, earning graduate degrees, paying bills and ordering pizza. By the mid-1990s, the Internet had stolen the thunder of the magic box. But since TV and computer screens were literally viewed differently—with users typically sitting 10 feet from the TV set and 18 inches from the computer screen—it was considered wise to pursue interactive content for both.

It was at the intersection of theory, engineering and hype that Malone shined. The magic box was 500 channels all over again. He had a purchase order from General Instrument, not for one million digital boxes to back up his boast, but 15 million! This consisted of

11.9 million that his company would order, plus 3.1 million by a consortium of nine other cable companies.

From his mainframe mind came the following download:

> Intel can now do MPEG-2 in software instead of in hardware. We don't have to invent HTML or Javascript or WebTV solo chips, or 200 MIPS processors. We don't have to write Windows CE. All of these things exist. So, it's more of an assembler operation than invention and creation. So, GI's got to use the scale of this purchase order to bang their vendors to get world class pricing and start down the road of silicon integration, so instead of ten chips, you end up with two. That's really the process of what drives costs down the scale ...

And so on.

The difference between 500 channels and the magic box is that this time Malone had two major software vendors, Sun Microsystems and Microsoft, eager to bang each other to get into his box. And there were retailers reportedly willing to subsidize the $300 per-box cost to get first crack at the viewers who would use it. Malone described, or rather sold, this potential participation "as the biggest bidding war in the history of the communications business."

The chess machine was ready to move the major pieces again and he would have a brand new headquarters from which to plot strategy. The building's circular design, said Malone, was to encourage a sharing of ideas. "Everybody should know as much as possible about what everybody else is doing." He said, half-seriously, that offices for him and Hindery would be in the back of the building, to avoid process servers delivering lawsuits. There would be underground parking and a perimeter security system and other protective features—just like the real Pentagon.

"The security issue has been of concern since the Unabomber sent me an e-mail a couple years ago," Malone said.

The Luddite terrorist sent a threat by e-mail?

That's what Malone said.

THE MAGNESS ESTATE feud began melting down in public just before Christmas of 1997. Court filings revealed that the executors, Donne Fisher and Dan Ritchie, had approached several companies about buying the disputed stock, including Disney, Oracle Corp. and US West, before selling it back to TCI and Malone. The Magness sons dismissed that attempt as "flipping through their Rolodex." The entire case, including Sharon Magness's claim for half the estate, was headed for a February 1998 court date in Colorado.

Stories in the Denver newspapers recounted the family feud and court battles, attention sure to draw coverage from national magazines and tabloid TV shows as the trial date drew near. When the story broke of Comcast's offer for the Magness estate shares—six months after the fact—the cable spin machine went into gear with a joint statement from Leo Hindery and Brian Roberts that read, in part, "We are greatly concerned by a report in today's *New York Post* of an alleged breakdown of the close relationship between our two companies. For many years, our companies have enjoyed a cherished and productive association, resulting in many key alliances and partnerships."

It was time to put the genie back in the bottle. Attorneys for TCI, the Magness sons and Sharon Magness worked through the Christmas holidays on a settlement proposal that was announced on January 5, 1998. The sons would retain half of the estate's supervoting B shares that had been pledged to TCI. Plus they would receive a payment of $124 million. Sharon Magness's settlement was confidential. Malone won the right to vote the Magness shares along with his own as a block, which gave him effective 40 percent voting control over TCI. He also received a $150 million payment from TCI, in exchange for the guarantee that his TCI supervoting shares would never be sold to an outsider, but be sold back to the company.

This payment raised some eyebrows, though it fit a long-established pattern at TCI of deals that benefited Malone. There was, however, a juxtaposition with Hindery's vocal criticism of programming and sports costs and how they affected subscriber rates. "Let's talk about Oprah Winfrey signing a $130 million TV contract and then we can talk about cable rates," he said when Kimmelman's report was issued.

Hindery was so apoplectic about sports programming costs that he was fined $100,000 by the National Basketball Association for criticizing Turner Sports for doubling its payment to retain NBA TV rights. TCI owned part of the New York Knicks and Ted Turner owned the Atlanta Hawks. Team owners in the clubby confines of the NBA are not allowed to trash-talk one another. The fine was rescinded, however, when Turner interceded with the league.

Appearing before a Senate subcommittee in October of 1997, Hindery had bashed the Minnesota Timberwolves for paying its young superstar, Kevin Garnett, $120 million to play basketball—a cost Hindery said would be borne largely by TV viewers. "I didn't give Mr. Garnett $120 million to play basketball for six years for the Timberwolves—my customers just did," he said.

Were TCI customers, therefore, responsible for the $150 million paid to Malone in the Magness settlement?

Hindery said no. Programming cost increases can pass directly to subscribers under FCC rules. The Malone payment was drawn from the increased stock value of TCI's companies, he said. But wasn't the value of cable stocks supported by subscriber payments, network launch fees and advertising revenues? Didn't the $150 million have to come from somewhere besides thin air? Or was that the greatest financial engineering feat of all?

—

IN THE LATE SUMMER of 1997, Hindery had met with Robert Allen, the chairman of AT&T, to discuss potential alliances, including the delivery of local phone service through TCI's cable lines. By November, Allen was history, replaced by C. Michael Armstrong. He came from GM Hughes with a mandate from the AT&T board to remake the world's largest and oldest phone company in his own image: a crusher.

Growing up in Detroit, Armstrong's athleticism won him a football scholarship at Ohio's Miami University. A shoulder injury put an end to his days on the gridiron, with the resulting surgery leaving his right arm less mobile than his left. With either limb, Armstrong looked like he could put the hurt on. At age 59, he retained the bulky physique and gregarious nature of a man used to running through human walls. From the neck up, he looked like Mr. Clean, his balding head like a shiny torpedo with laser-guided eyes.

He worked 31 years at IBM, rising through the ranks to head its international division before realizing he would not get the top job. He moved to GM Hughes in 1992, where he oversaw a remarkable turnaround. DirecTV was already in process by the time he got there, but he ran blocker with the GM board and oversaw a satellite TV service that captured four million subscribers in four years. He restructured the broader Hughes satellite and communications businesses, cutting 25 percent of its workforce and selling its defense business to Raytheon.

GM Hughes stock rose from $20 to $67 during his tenure.

Those were the kind of numbers AT&T wanted to see. Armstrong got the nod. One of his first orders of business was to visit Malone, whom he had met with eight months before when their mutual trouble was Murdoch and Ergen's Deathstar. It seemed like ancient history. And in the time it takes to play cutthroat, it was.

One of AT&T's major problems was that long-distance phone service had become a commodity market, with all the arbitrage of prices that implies. It was not just MCI and Sprint as competitors, but no-name dial-around services. Old-timers who grew up with Ma Bell and a leased black phone were suddenly taking their long-distance discounts right at the phone keypad, gabbing for nine cents a minute or less with dial-around services. No need to pay 25 cents a minute to subsidize the long-distance companies' expense accounts and celebrity pitch people.

Meanwhile, competitive access providers (CAPs) were building fiber-optic loops around metro areas to cherry-pick long-distance business customers. And a technology called packet switching, which carries Internet traffic, promised to be the Deathstar of phone pricing. Traditional telephone calls require an open line from one caller to another. Packet switching can carry voice traffic over the Internet: the world is a local call. Soothsayers predicted free phone calls within a decade. Phone companies were going to have to find new revenue sources to survive.

These economics were particularly vexing for long-distance companies because one-third of their revenues—more than $30 billion a year—went to the local phone companies in fees to switch long-distance calls to their endpoints. The Baby Bells were the trolls under the bridge.

That is some of the competitive landscape that Armstrong faced at AT&T. He immediately announced plans to cut 18,000 jobs. He gained a $4 billion war chest and wiped $5.6 billion in debt off AT&T's books with the sale of the Universal Card and another business unit to Citicorp. He dissolved an investment he helped create: selling back the 2.5 percent stake of DirecTV that AT&T bought in 1996.

Armstrong also took a look at a CAP named Teleport, which was founded in 1983 by Robert Annunziata, a former AT&T executive. It targeted business customers in 65 cities, including New York and Los Angeles. TCI, Comcast and Cox were part-owners of the company and helped build its network in major metro areas. Before joining Teleport, the cable gang's fledgling phone network was a way for AT&T to play cutthroat with the Baby Bells. Here's how it worked, according to Malone:

> AT&T was in here all the time, telling us where we ought to build, where their traffic volumes were, where they could give us business and so on. They were stimulating the cable industry to do that, but they weren't ready to invest in it because they wanted to trade the cable industry off against the [Baby Bells] if they could get a big switch order out of Ameritech or they could get a lower access fee out of US West. Then they would come back to the cable guys and say, "Sorry about that area, you know, build somewhere else." So we were getting traded out and we knew it and weren't too thrilled with it.

The payday came when Armstrong concluded that AT&T should buy Teleport outright, a deal announced on January 8, 1998, three days after the Magness estate litigation was settled. AT&T acquired Teleport for $12.9 billion in stock, which translated to almost 47 million AT&T shares for TCI in exchange for its 30 percent Teleport stake. Suddenly, Malone held a huge stake in AT&T, just as he did in Time Warner, in the @Home cable Internet service, in Discovery Communications and Encore Media and Black Entertainment Television and Primestar, plus half of Fox Sports.

The chess machine reigned ubiquitous.

———

THE WORLD MOVES on, but Dan Garner did not.

He could not accept that he had lost the 110 DBS slot and the hundreds of millions of dollars it came to represent. After exhausting his appeal of the FCC auction decision all the way to the U.S. Supreme Court, where he lost in 1997, Garner tried a civil suit. He filed against MCI in Arkansas state court on February 20, 1998, seeking damages of $735 million, the amount that the 110 and 148 slots brought at auction in 1996.

The suit claimed that MCI's written offer in 1995 to submit an opening bid of $175 million for the 110 slot unduly influenced FCC chairman Reed Hundt to cast the deciding vote for auction.

MCI vigorously denied Garner's suit, sought to have it dismissed and claimed that Advanced "apparently sees in MCI a deep pocket from which to recoup its losses."

Garner's attorney, Robert Ginnaven, said that the underlying facts of the case had never been tried and that he would seek to depose the whole cast of characters from that time, including Hundt and MCI chairman Bert Roberts.

Assuming it ever got to trial, the questions raised would be: Who really owned the 110 slot? Who had the authority to sell it? and What was the motivation for auctioning it?

Hundt, working as a telecommunications consultant in the private sector, said in an October 1998 interview, "The Advanced case was about a slot that should be auctioned. Everything else is stuff made up by lobbyists. The slot belonged to the public. It was auctioned. That's what we should do with all spectrum. It went to the person who paid the most."

Garner's case, which appeared to be a long shot, was transferred to federal court jurisdiction in Arkansas. Three separate federal judges were obligated to recuse themselves from considering the case because they owned stock in MCI, which became MCI World-Com after a 1998 merger.

As of June 1999, Dan Garner was still chasing the ghost of 110 through the courts.

—

A WEEK AFTER Garner's suit was filed in Arkansas in 1998, Michael Milken settled a claim with the U.S. Securities and Exchange Commission alleging that he violated his lifetime ban from the securities industry by advising on two deals, including the MCI/News Corp. alliance in 1995. Milken's company, MC Group, made $27 million for advising Murdoch and Roberts. There was a $15 million fee from another deal involving News Corp. and $5 million interest accrued on both. Recovering from prostate cancer and fearing a return to prison if a violation of his probation was determined by a court, Milken agreed to settle the case by depositing the $47 million in fees into the U.S. Treasury. Michael Milken was free to go.

He had, in fact, already moved on to create a company named Knowledge Universe that pursued a range of educational technology investments. A Milken-led group, including TCI, made an unsuccessful bid for the educational publishing arm of Simon & Schuster in 1998.

Milken's most enduring legacy, the junk bond, not only survived but thrived in the booming U.S. economy of the 1990s. There were $119 billion in high-interest yield bonds issued in 1997, triple the amount sold in 1986 when Milken was riding high at Drexel Burnham Lambert and began to be vilified as all that was wrong with America.

If the free market is a reliable indicator—including the ascension of companies like MCI, News Corp., Turner Broadcasting and TCI—then there is much that is right with junk bonds. Though Milken was barred from the action, there were plenty of entrepreneurs—Charlie Ergen among them—who were telling their business dreams to a new crop of junk bond dealers.

⁓

ERGEN PADDLED HARD to keep his company afloat. EchoStar broke the one million subscriber mark in December of 1997 and even poked some fun at Murdoch while the company's $5 billion breach-of-contract lawsuit against News Corp. worked its way through the courts. A holiday television ad featured Santa considering who was naughty and nice. An elf suggests "J. R. Murdock," setting off a red warning light that flashes the word "Naughty."

Santa says, "Tsk. Tsk. Let's give him … cable!"

EchoStar remained virtually a one-man show, with Ergen as Elvis.

John Reardon, a former president of MTV who later worked at TCI and was dismissed in the Hindery purge, met with Ergen in late 1997 to discuss joining EchoStar as a senior executive. Having heard the combat stories from departed EchoStar executives, Reardon put on a presentation he called "Good Charlie, Bad Charlie."

"I told him if he really wanted to run a $1 billion company, he was going to have to change the organizational structure," recalled Reardon.

He joined EchoStar in January of 1998 as president of Dish Network.

Eight months later, Reardon was gone, head spinning from the pace, the long hours and what he described as Ergen "pounding people into submission" to pursue company goals. "I've worked with Sumner Redstone, Steve Ross, John Malone and Peter Barton and I have never experienced anything like this guy," Reardon said.

Reardon ran into Carl Vogel at an MTV event in Los Angeles later in 1998. They gave each other big smiles, as befits two veterans of the Elvis rock-and-roll revue.

"So, you put up with him for three years," Reardon said.

"Charlie's problem is he believes his own shit," said Vogel.

—

IN MAY OF 1998, Bill Gates addressed the National Cable Television Association convention in Atlanta. He had struck a deal with Malone to put the Windows CE operating system into five million of the magic boxes. Gates's rival, Scott McNealy at Sun Microsystems, had a similar deal for his Java software. Malone, the master of leverage, had played the two software chiefs against one another. During the tense last stage of negotiations, Gates was on a flight, talking by airphone with Malone on the ground. "You're rocking, Bill," Malone said, needling the younger man for his nervous tic.

In his address to the cable executives in Atlanta, Gates projected on screen his own mug shot, taken after his 1977 arrest for what he said was a speeding violation in Albuquerque, New Mexico, where Microsoft was founded before moving to the Seattle area. Gates tried to warm up the crowd to embrace Microsoft's cable ventures. "There's no need for paranoia," he said. "In fact, the cable industry is a confident industry and that makes it a pleasure to work with."

Talking to reporters afterward, Malone said, "A good friend of Bill's, [Intel's] Andy Grove, says only the paranoid survive."

—

THE PRIMESTAR PARTNERS were, as always, in a state of flux. Preparing to roll up into a public company with Rupert Murdoch on board, the cable gang pulled out $479 million from the company in early 1998. This seemed odd. After all, if they wanted to create a vibrant new satellite TV service, with 18-inch dishes and 200 channels, it seemed that investing more money in the business was the way to go. The gang decided, however, that they would take the money out and replenish it by raising $400 million worth of debt under the banner of Primestar Inc.

This assumed that their deal with Murdoch would pass muster with the U.S. Justice Department and its antitrust chief, Joel Klein.

He grew up in a public housing project in Astoria, Queens, and graduated summa cum laude from Harvard Law School. He clerked for Supreme Court Justice Lewis Powell and became an appellate judge. Klein served as deputy White House counsel before joining the Justice Department as assistant U.S. attorney for antitrust in 1996.

Klein was criticized by those who feared increased media concentration, after he signed off on the merger of Bell Atlantic and Nynex in 1997. Klein also inherited the government's antitrust case against Microsoft, which was headed for trial. Klein said his preferred method of antitrust action was "surgical intervention."

Primestar was quite a corpus to operate on. In terms of financial size, the $1.1 billion deal with Murdoch was almost negligible in the crush of mergers and buyouts. But this was a high-profile case, given Primestar's tortured history, the cross-ownership issues and the well-documented Rupe-a-dupe of Ergen. Whether or not Murdoch had breached his contract with EchoStar was a matter of litigation for those two parties.

The strict legal question for Justice was whether the reconstituted Primestar would help or harm competition in the marketplace of "multichannel video programming services," a market that broke down to 87 percent of subscribers served by all cable companies, 9 percent by satellite TV and the rest by other competitors.

One measure of market concentration is the Herfindahl-Hirschman Index, calculated by squaring the market share of producers and adding it up. An index of 10,000 is a pure monopoly, with numbers ranging down from there. The Herfindahl-Hirschman Index is a thumbnail calculation that did not bode well for the cable gang aligning with Murdoch to launch a rat zapper.

It was the Justice Department depositions of Murdoch, Malone and Hindery, however, that sealed the deal's fate, revealing just how shaken the cable gang was by the Deathstar deal and what TCI's hitmen did to stop it.

The deal "had a chilling effect on the stock market for cable," testified Malone, whose 1997 interview with *Satellite Business News* was also quoted, in which he said, "There's some kind of peace in which Rupert gets what he wants, which is broader distribution of his programming networks in exchange for which he's not quite as aggressive" in satellite TV.

Murdoch sought advice from Malone, who told him that his plan to take on cable by satellite was "lunacy." To which Murdoch replied, "Well, then, help me get out of it, help me find something else to do ... what is Plan B?"

Plan B was Primestar.

Discussions were held between Chase Carey, News Corp.'s chief operating officer, and Hindery, TCI's president, who negotiated on Primestar's behalf. Hindery's role was one of "a peacemaker. ... He kept trying to convince everybody that there was more profit in peace than war," said Malone.

Reaching an agreement among Murdoch and the Primestar partners would "serve all masters well to the benefit of all," Hindery testified.

All masters, that is, except for consumers seeking competition with cable.

On May 12, 1998, the Justice Department filed to block the transfer of the 110 slot and related satellite TV assets between News Corp. and Primestar. The announcement came just as Primestar officials were on a road show, trying to raise $400 million from investors. The road show was canceled.

"It was a little bit of a Pearl Harbor," Malone said of the Justice Department's timing.

The government's lawsuit dangled a possibility that Primestar's deal with Murdoch might be allowed if the cable companies divested their ownership.

That suggestion was about eight years too late. The Justice Department, under different leadership, had a chance to reach that conclusion in 1990 when they began their initial antitrust investigation into Primestar. Instead, they approved a 1993 consent decree that allowed the cable gang to maintain ownership. Then came Malone's attempt to buy the 110 slot from Dan Garner, the machinations of Reed Hundt's FCC, the auction of the slot won by MCI/News Corp., Murdoch's Deathstar deal with Ergen, Malone's clampdown and, finally, the Justice Department's lawsuit to block Murdoch's deal with Primestar.

By focusing so narrowly on each issue as it arose, the Justice Department, the FCC and Congress missed the forest for the trees. The chess machine's game plan had become increasingly clear from the mid-1980s on: maintain cable's monopoly, prevent programmers from independently aligning with satellite TV services, launch Primestar to mop up the uncabled market, fog the future with 500 channels and gain control of rat-zapper spectrum either by purchase or obstruction. And it was all perfectly legal.

The chess game between Malone and the federal government was the most interesting one of all. The government's side was played by an ever-changing cast: First it was Al Gore in Congress, then antitrust chief James Rill at the Justice Department, then Hundt at the FCC, then Klein at Justice. Malone remained steadfast and unchanging, his eyes on the prize, which was not necessarily to win this game, but to buy time so that he could play on all the other chessboards of cable system ownership, programming, technology and Internet investments.

The rat-zapper endgame had arrived. Malone was in checkmate. But he was already gone.

—

GLENN JONES ENJOYED himself thoroughly.

He and 70 other people sat in a small theater in Denver as a troupe of cable industry workers performed a satirical musical revue to benefit the industry's AIDS organization, Cable Positive. The show's lyrics were penned by Erica Stull, an executive of Jones Intercable, which Glenn Jones founded in 1970.

The performers hopped around the stage, singing versions of Broadway show tunes such as "If I Had a Website," "Hello, Leo" and "Everyone's Putting Up Dishes," which featured these lines: "Aim at that satellite / All you need is a good line of sight / You'll be swell, even so / Trim those trees, clear that snow / And don't call Charlie Ergen 'til you do."

Glenn Jones, 68, was a bit of a poet himself and a self-styled Renaissance man for the Information Age. He had been a Navy frog diver, played the piano and read widely, particularly futurist tracts by Isaac Asimov and others. One of his nicknames was Galactic Glenn, because the cable business was his platform to pursue the future of technology and education. He created Mind Extension University, which allowed students to take courses and earn degrees by accessing a cable TV channel and Internet connection as a virtual classroom. "All of this technology terminates at a three-pound electro-chemical device inside the head," Jones was fond of saying, meaning the brain.

Jones made $60 million selling 30 percent of his cable company to Bell Canada in 1994 amid the rush of phone company interest in cable companies. Bell Canada had an option to buy him out completely by 2002, but was disappointed with the cable company's performance. Bell Canada was also feuding with Jones in court over ownership of Internet services deployed over cable lines.

One of Jones's sidelines was publishing what he called "Briefcase Poetry," paperback books he wrote under the pen name Yankee Jones. A poem titled "Business Advice" consists of just two lines: If you live with sharks / Don't bleed.

Good advice, especially since Jones was about to get eaten—by Comcast.

Over the Memorial Day weekend of 1998, Comcast chairman Ralph Roberts flew out to Denver to tell Jones that Comcast had bought Bell Canada's stake in his cable company and related options for $500 million and was interested in acquiring the rest.

Glenn Jones's endgame move was made for him. The deal closed in September of 1998, and Jones received $200 million to cash out his ownership. Comcast added nearly one million subscribers to its portfolio of cable systems, which already served 4.5 million.

The investment by Microsoft in Comcast in mid-1997 had lit the fuse of cable stocks and speculative investments. A new player at the cable table also had a Microsoft connection: Paul Allen, a co-founder of the company who left in 1983, but remained a board member and friend of Gates.

Allen's Vulcan Ventures invested widely in technology and entertainment properties. He owned 80 percent of Ticketmaster, later sold; a stake in Dreamworks SKG, the movie studio led by Steven Spielberg; 100 percent of the Portland Trailblazers basketball team; and stakes in the Starwave and CNET Internet services.

Allen spoke of a Wired World, marrying content to computing power, similar to Gates's concept of the Web lifestyle. In the early and mid-1990s, Allen and his investment adviser, William Savoy, made several investments in rat-zapper technology. They bought into SkyPix, the ill-fated DBS service. Allen also bought 4 percent of United States Satellite Broadcasting, the partner of DirecTV. Allen was techie enough that when he visited the USSB plants in Minneapolis and Indianapolis in the summer of 1994, he opened the decoder boxes to check out the circuitry. Allen also reportedly owned a stake in EchoStar in early 1997. There was potential for more, but Murdoch swooped in and set the Deathstar deal in motion.

Whatever Allen's interest in satellite TV, it was left in the shadows by the huge move he made into cable in 1998. In April, he spent $2.8 billion for Marcus Cable, the Dallas-based company with 1.2 million subscribers. In July, Allen struck again, with a $4.5 billion deal for Charter Communications of St. Louis and its 1.2 million subscribers.

Allen paid $3,750 per subscriber for Charter, almost double the industry standard of $2,000. The day of the Greater Fool had arrived. Allen was not only willing to pay the private market value by which cable companies sold each other properties, he was paying a big premium. Whether he overpaid, or got a bargain because the cable pipe is a guaranteed cash gusher for the Wired World, doesn't really matter because what something is worth is what someone is willing to pay for it.

It all depends on the pitch. And Paul Allen had a lot of money to pitch. In the course of the next year, he would spend billions of dollars for cable companies that represented five million subscribers and counting. He had bought himself a seat at the cable table.

—

IN EARLY JUNE, TCI announced that its United Video Satellite Group would pay $2 billion to buy News Corp.'s *TV Guide* and related assets. The merger allowed Murdoch and Malone to control, and expand, the Prevue electronic programming guide into the digital future and perhaps keep Gates at bay. TCI would own 44 percent of the merged entity, with News Corp. holding 40 percent. The EPG and the company itself would be renamed TV Guide. To clear away some pesky print competitors, News Corp. bought TVSM Inc., publisher of *Total TV* and *The Cable Guide*, for $75 million and tossed them into the mix.

This corporate cluster then made a hostile bid for the only other viable EPG competitor, Gemstar, which owned the StarSight technology. The companies had been battling in court over patent issues. But Gemstar's owners wouldn't sell out for $2.8 billion. They didn't understand the power of aligning assets to become bigger and bigger and bigger. Or, they just wanted to remain independent—the most risky proposition of all on the interlocked electronic frontier.

—

ON JUNE 24, 1998, TCI was sold to AT&T in a $48 billion deal.

Malone would become AT&T's single biggest individual shareholder and run the Liberty Media programming company, which remained an independent entity under AT&T. Leo Hindery became president of the newly named AT&T Broadband and Internet Services, a mouthful that encompassed AT&T's intention to provide combined TV, Internet and local phone service over cable lines.

Malone once said that he was viewed early in his cable career as a "Manchurian candidate," sent by his old employer, Ma Bell, to infiltrate the cable industry and take it over from within. That turned out to be close to the truth. Malone's sale of TCI triggered a wave of cable company sales and consolidation. Malone had seen that coming on a wide scale, across industries. "Two or three companies will eventually dominate the delivery of telecommunications services over information highways worldwide," he said in 1994. "The big bubbles get bigger and the little bubbles disappear."

AT&T's purchase of TCI provoked a wide range of opinion and questions. Some saw it as bringing the competition promised in the 1996 Telecommunications Act, since AT&T vowed to compete with the Baby Bells. But would the telecom animals really scratch and claw and bleed, or would they pose? How many billions of dollars would it take to deploy new services? What was the business model for recapturing that investment? How much would consumers be willing to pay and what would they get in return?

In Milford, Connecticut, patrons of the River 3 barbershop were unaware that this landmark deal in the communications world came from someone, John Malone, who grew up in their town. But they were skeptical of what was to come. These deals "are only good for the investors and the shareholders," said Jim Bonnanzio, an unemployed shipping industry worker and father of three, awaiting a haircut. "It's money in the bank for them, but how much more service can you give? They want to put the Internet on TV. The kids don't even know how to use the Dewey Decimal System at the library."

Fifteen miles away, at the Hopkins School in New Haven, construction was ongoing for the Malone Science Center, the facility that John Malone had funded with a $5 million gift in honor of his father, Daniel. John Malone said that the bulk of his wealth will go toward education, funded by his Malone Family Foundation. He said he did not want to leave much to his two adult children because it might rob them of their initiative.

Regardless of what he planned to do with his money, the way that he earned it at TCI left some observers with a bad taste.

James J. Cramer, the frenetic mutual fund manager and business columnist, weighed in heavily on TheStreet.com Internet site after the AT&T/TCI deal hit the news. "Malone piles on debt," wrote

Cramer. "Malone pisses off consumers. Malone stands for anti-consumer, pro-monopoly, high-priced, anti-technology answers to pro-consumer questions. Malone is an '80s guy, trapped into thinking how many homes he can hoodwink. Innovation. Forget about it."

Malone was a magnet for that type of criticism his entire career. Was it true? Malone addressed the subject in an interview shortly after the AT&T/TCI deal was announced.

"TCI has made most of its wealth for its shareholders through investment and financial engineering. Okay? It was for years, you know, a substantial cable operator and that gave us the platform with which to do all this other stuff, but that's really the business we've been in. I'm not going to claim that we've always been the greatest cable operators in the world. But we've been the smartest guys in the industry in terms of building wealth."

In the cable industry, and every other business, companies large and small may strive to be the greatest—out of pride or morals, or because it makes the cash register ring.

That was not the Malone way. He looked at the world without sentimentality. Being the greatest general store may be a worthy goal, but how many of those stores were sunk by the arrival of Wal-Mart on the outskirts of town? How many subscribers who never had a major problem with their cable company bought a satellite TV dish anyway? How many bookstores lose business to Amazon.com or its equivalent in cyberspace? The free market is free because it swings between seller and buyer and back.

The chess machine was the unalloyed avatar of capitalism. Why didn't subscribers get 500 cable channels or the magic box or any of the other promised goodies that spilled so easily from Malone's lips? A generous interpretation comes from Royal Little, the father of the modern conglomerate, Textron, which owned cable systems in the 1970s. "Never let an inventor run a company," he once said. "You can never get him to stop tinkering and bringing something to market."

That was one aspect to Malone, the tinkerer, but there was another, the steely-eyed pragmatist with the perfect pitch. Those who bought his blue sky of 500 channels and the magic box should know that the man who painted it stuck with the wisdom of the market. "You get more bang for the buck if you're late in the game," he once said. "That's the technology cycle. So you really don't want to spend your money too early or you have too early obsolescence. The other thing is you don't want to be too late, or you lose market share to competition. So, you're trying to find that balance as you go along."

That's why TCI systems typically offered fewer channels than other cable companies and delayed deploying new services. From Malone's view, if the demand was there to improve, he would supply it. If not, then not.

That's been his game.

Some visionaries come from the school "Build it and they will come." Malone's school was "I'll promise to build it and then let's see what happens."

Monopoly local franchises gave him the cover to operate that way, but it is the government's job to set the rules. Malone played it as it lays. His laissez-faire style nearly capsized TCI in late 1996 and early 1997. Subscribers were bailing, the debt pyramid was toppling and Deathstar was on the horizon.

Malone reacted by bringing Hindery in to clean up the place and get it ready for sale, while he bought time by keeping Deathstar at bay. Three variables on the electronic frontier are money, technology and leverage. All these can be acquired. The constant is time. Buying time is the hardest—it's the fourth dimension of the game. Malone could play that, too.

The investment and deployment required to deliver new gadgetry and digital services into every home—by cable, phone line, wireless or satellite—require scale economics and a willingness by businesses and consumers to pay for it all on the installment plan.

By jerks and spurts, that future will arrive in the twenty-first century just as an earlier version did in the twentieth century. Malone's grand vision will someday be confirmed. Like John D. Rockefeller in the oil business, Malone is a complex figure—the kind needed to engineer an industry, but not squelch competition; the kind needed to create wealth by building assets, but not responsible for serving the people who foot the bills.

Having an architect as a service station attendant is asking for trouble.

Malone found plenty of it. His clashes with government officials such as Gore and Hundt were ideological and raised questions that go to the core of a civil society: What is the responsibility of business? What is the proper role of government?

The irony of Malone's railing against official Washington was that his financial engineering resembled its budget methods. Cable rates (taxes) went up every year to fund programming (services) and pay interest on the ever-ballooning debt (debt), while the bottom line often bled red (deficit). Malone's endgame required a buyer. It is unclear what the federal government requires.

It may take years for Malone's legacy in American business to become clear, when there is more light than heat. But there is one aspect that is certain. When the J. D. Power and Associates survey of customer satisfaction among cable and satellite TV firms came out in September of 1998, TCI was dead last among 14 providers for the second year in a row, though it had improved somewhat under Hindery's reign.

In March of 1999, however, a *Fortune* magazine survey of executives, boards of directors and securities analysts selected TCI as the most admired cable company in the country, based on a variety of business measures. It was not even in the top 10 the year before.

Malone was proud of that accolade. Absent from his pride was the recognition that he might have had it all—financial success for investors and himself, along with the admiration of a job well done, starting with Vail in 1973 and ending with the sale of TCI to AT&T a quarter century later. The chess machine could have been the "greatest cable operator in the world" if he set his mind to it.

He just didn't want to.

IN JUNE OF 1998, EchoStar held a nationwide sales meeting at the company's new headquarters, a onetime shopping mall south of Denver that the company had purchased out of bankruptcy and refurbished. Several hundred dealers from around the country ate hot dogs, drank beer and talked shop. Ergen sat in a second-floor conference room with a reporter. He sipped a beer and recalled the events of the past 16 months: from the deal with Murdoch that lit up the sky, then fell to earth, and the slow road back.

"We were left for dead," he said. "We were left by the side of the gutter for dead by News Corp. and TCI and Primestar. I don't think anyone believed we'd survive."

As he spoke, EchoStar's $5 billion breach-of-contract lawsuit against News Corp. was moving through federal court in Denver. News Corp. had countersued, alleging that Ergen negotiated the original merger deal in bad faith.

EchoStar attorneys had subpoenaed documents from every one of Primestar's cable partners and News Corp., and began videotaping depositions of the principal players. The Justice Department's lawsuit against Primestar appeared to bolster EchoStar's case, which made sense, because EchoStar gave federal investigators as much information as possible to help them make it.

At headquarters, Ergen psyched himself up with the EchoStar underdog theory.

"I think we're going to trial," he said. "I don't think there's anything they can offer us. Howard Hughes said everyone has their price and they assume that and run their business that way. And I think it's a miscalculation as it's related to EchoStar. It's not about the price. It's about the way big business works and about making that public and letting somebody decide whether we were injured or not. Every American can identify with fighting the big guy."

—

THE ODDS OF A high-profile business lawsuit going to trial are slim because of the time and expense and the uncertainty of the outcome. Of all the business battles and lawsuits and cutthroat strategies that result in injuries real and feigned, few cases end up before a judge or jury. One exception was US West's lawsuit against its business partner, Time Warner, to block the acquisition of Turner Broadcasting System in 1995. The merger went ahead, a Delaware Chancery Court judge ruled against US West and the partners settled their differences. Evident at the trial, however, was the stark fear with which the captains of media companies view each other. Like pirates on the high seas, they know not whom to trust nor whom it is safe to ally with in the mutual pursuit of plunder. So while they disagree on strategy, stab each other in the back and go to court, they must remember that tomorrow is another day and they will switch sides and do it all over again.

It came as no great surprise, therefore, that on November 30, 1998, EchoStar and News Corp. announced they were settling their litigation. In exchange for dropping its lawsuit, EchoStar would get the 110 satellite slot, two satellites launched into orbit and the uplink facility in Arizona. EchoStar would also manufacture 500,000 set-top boxes for international sale using the News Corp. encryption system, which had been a sticking point in the earlier deal.

News Corp. and MCI WorldCom would get EchoStar stock valued at $1.2 billion, which amounted to 15 percent of the company. The total voting power by News Corp. and MCI was only 3 percent, however, and they would have no operating role in the company going forward and no board seats. Ergen, as he had from the beginning, retained 90 percent of the voting power in his company.

It was like drawing four threes in a game of seven-card stud.

On a press phone call to announce the settlement were Ergen and other EchoStar officials, as well as Chase Carey, the News Corp. chief operating officer who had asked for Ergen's resignation 19 months earlier in the heat of the battle.

"I don't think that statement is accurate," said Carey when reminded of the allegation in EchoStar's lawsuit. "Mr. Ergen is running the company and we're going forward."

It was as if none of it had ever happened, as if News Corp. and EchoStar were just this day deciding to join forces in the satellite TV business. "It's a situation that I think is a win/win," said Ergen, as Carey chimed in with some other business-speak.

Deathstar lives!

But not really. Ergen finally had what he wanted to go after the cable gang. He quickly received approvals from the U.S. Justice Department and the FCC to use the 110 slot, but it would not be the same as if Murdoch were on board in 1997. Two years was an eternity on the electronic frontier and it remained to be seen what kind of star was EchoStar.

——

CARL VOGEL SAT in his office three days before Christmas of 1998 and contemplated his fate.

After the Justice Department case was filed against Primestar and its cable owners, Vogel was brought in to straighten out the mess and get the company back on track. He learned quickly what Primestar's former president, John Cusick, had found out: He headed a company at cross-purposes.

Murdoch and Malone had tried to cut a deal, to find a way around the Justice Department lawsuit, perhaps by buying out the other Primestar cable partners. But the parties could not come to terms. Outside investors who took a look at Primestar were skittish when they saw that the cable gang had pulled out $479 million from the business earlier in the year.

Primestar could have challenged the Justice Department lawsuit and tried to keep the 110 slot. But a trial would have begun in February of 1999, a month before the cable industry was to be released from the yoke of federal rate regulation. Going to trial would raise the profile of the cable gang and provoke questions about the state of competition between satellite TV and cable.

Without the full 110 slot, without the high-powered rat zapper, Primestar was doomed. It had two million customers, but lost almost as many as it gained each month. Both DirecTV and EchoStar were offering dealer bounties to convert Primestar customers. Primestar had access to the DBS bird launched by TCI/Tempo to the partial 119 slot in 1997. Yet, Vogel would need to get the myriad Primestar partners to support and fund a new DBS service limited to 100 channels. That idea was going nowhere.

Primestar was at a dead end and Vogel knew it. Just as in the cable industry, the shakeout and consolidation happened fast in the satellite TV industry. Opportunities delayed were opportunities lost. The only scenario that made sense was to sell off Primestar, its subscriber base, the TCI/Tempo bird and another Loral satellite in storage. The buyer, at a total price of $1.8 billion, would be DirecTV, which had already announced plans to buy out its partner, United States Satellite Broadcasting, for $1.3 billion.

With acquisitions and growth, DirecTV would break the eight million subscriber mark. It would later receive a $1.5 billion investment from AOL to pursue the delivery of Internet services to homes by satellite. DirecTV was a dominant force in satellite TV that Primestar might have become.

Vogel had lived a few lifetimes in the business, though he was just 41 and looked younger, with a cherubic face and a curly crown of brown hair. He had worked 11 years for Glenn Jones, rising to a top spot. He joined EchoStar just as they were beginning to take off and spoke wistfully of that time. "We were on the edge a lot," he said.

Carl Vogel sat in his chair, waiting to make his next move. On his left wrist was a black band that one of his five children had given him as a reminder. There were four white letters on the band that spoke a silent question—one that had no place in cutthroat.

W.W.J.D.

What Would Jesus Do?

EPILOGUE

SURVIVAL OF
THE LEVERAGED

T HE U.S. ECONOMY cruised into mid-1999 in overdrive. The Dow Jones Industrial Average broke 10,000 points. Unemployment was low and inflation was modest. The Bill Gates personal wealth clock ticked toward $100 billion.

Where Internet stocks were concerned, Wall Street and investors continued valuing initial public offerings of companies with no earnings on very little but the prices of other Internet companies with no earnings.

"It's the Wall Street way. We're all greedy as hell," said Edmund Cashman, head of the syndicate desk at Legg Mason Wood Walker.

In a time of unprecedented domestic prosperity yet millennial anxiety, there were some constants to rely upon, like Rupert Murdoch dodging the taxman. An American citizen, he accomplished this feat with offshore subsidiaries and creative accounting. The company's effective tax rate worldwide was 6 percent, compared to Disney's 31 percent. There was one stiff levy Murdoch would not avoid, however: the divorce tax.

He and his wife, Anna, separated after 31 years in 1998, an event widely reported, including in Murdoch's *New York Post.* Ever vigilant, Ted Turner offered his views. Anna Murdoch is "a pretty smart woman," he said. As for his nemesis, "67 is no time to be dating, I'll tell you."

Anna Murdoch filed for divorce in Los Angeles and later stepped down from the News Corp. board. Her attorneys began combing over the Murdoch family's $12.7 billion trust, which included a 30.1 percent stake in News Corp. The couple had set up the trust for their three children, all of whom worked in executive positions in the company. Lachlan, 31, appeared to be the heir.

Rupe-a-dupe landed at the trendy Mercer Hotel in New York's Soho district, ensconced with Wendi Deng, a 31-year-old vice president of Star TV, Murdoch's Hong Kong–based satellite TV service. The mogul's wanderlust was gossiped about and publicized, just as his tabloid newspapers and Fox shows have fed on the sex lives and dirty linen of others, rich and poor.

"Rupert Murdoch is feeling the chill of the earth six feet down," wrote a columnist in *The Guardian* newspaper. In November of 1998, Murdoch returned to Adelaide, Australia, to attend the funeral of his first wife, Patricia Maeder, who died at age 70. Murdoch read a passage from the New Testament during the service.

An initial public offering for Fox Entertainment Group raised $3 billion. Murdoch furthered his reputation as the most powerful, and controversial, figure in sports, buying the Los Angeles Dodgers for $350 million and failing to buy the Manchester United soccer team for $1 billion.

As in the past, Murdoch shaped his business to suit his politics. Fox News gave cybergossip Matt Drudge a platform to muck around in President Clinton's affairs. Yet, Murdoch nixed plans by Fox TV to make a movie about the circus surrounding the 1991 nomination of conservative Clarence Thomas to the U.S. Supreme Court.

And, Murdoch shaped his politics to suit his business.

In 1993, he told News Corp. advertisers in London the simple truth: "Advances in the technology of telecommunications have proved an unambiguous threat to totalitarian regimes everywhere. Fax machines enable dissidents to bypass state-controlled print media; direct-dial telephony makes it difficult for a state to control interpersonal voice communications. And satellite broadcasting makes it possible for information-hungry residents of many societies to bypass state-controlled television channels."

The remark provoked the Chinese government to ban private ownership of satellite dishes and stiff-arm Murdoch from expanding Star TV into mainland China. He rehabilitated himself, first by yanking the BBC off Star TV to appease Chinese leaders. "We're not proud of that decision," he said later. "It was the only way."

Then, a year after China reclaimed Hong Kong from British rule in 1997, Murdoch reportedly spiked a book by Chris Patten, the former Hong Kong governor, due for publication by News Corp.'s HarperCollins. It was published elsewhere.

Murdoch made trips to Beijing and paid for Chinese officials to visit Britain. News Corp. contributed $1 million after flooding devastated China. In December of 1998, after five years of repentance for offending the Communist regime, Murdoch was embraced by China's president Jiang Zemin at a state guest house. Jiang praised Murdoch for his "objectivity" and signaled that News Corp. could gingerly pursue China's 1.3 billion potential TV viewers.

Murdoch's contortions describe a man who lives to undercut the established order—until it's time to cut a deal. He was that way at the beginning of his career and he was that way at the end. At age 22, Murdoch became publisher of the struggling Adelaide *News* and Adelaide *Sunday Mail* in 1953, after his father died. The much larger Herald Group sent a letter to Murdoch's mother, Elisabeth, threatening to launch a Sunday newspaper in Adelaide unless she sold them the *Mail*.

"I then did something that I like to think set the tone for the behavior of our company," Murdoch recalled 45 years later at a News Corp. summit in Sun Valley, Idaho. "I broke the rules of the establishment and published our opponents' offer on the front page under a headline that screamed, BID FOR PRESS MONOPOLY! And I included in the story a photograph of the confidential letter to my mother. That ruined any chance I might have had of being invited into the better clubs of Adelaide."

One is left with the legend of Murdoch vanquishing the competition, sacrificing social position and laying the philosophical cornerstone for the News Corp. empire. The truth is more telling. His opponent did indeed launch a rival Sunday paper named the *Advertiser*. After a brutal circulation war of several years, the two sides agreed to form a single Sunday newspaper. Murdoch controlled the resulting entity, a monopoly achieved through conflict, then compromise. His attack on the U.S. cable gang followed the same pattern. "Those cable guys are greedy," he told Ergen. "They don't want to put my programming on. I'll show them. I'll bring them to their knees." But when Murdoch was knocked to his own knees, he cut a deal—several of them—to get out of it. Murdoch's business interests were so enmeshed with Malone and the cable gang that he could no longer act independently. His ambition was bold, but his hands were tied.

Plus, he had a broader agenda. He wanted to buy The Family Channel from Pat Robertson to transform it into the Fox Family Channel. He wanted Fox News launched by Time Warner. He wanted Fox Sports to expand and thrive. He wanted to get into the electronic programming guide business. He did all of those things.

So, was Murdoch's gambit with Ergen against the cable gang foolhardy because it failed, or brilliant because it gave him the leverage to succeed elsewhere? He knew that the Sky deal would be provocative. But that provocation turned out to be the lever by which Murdoch fulfilled his other goals. He even ended up with 12 pecent of EchoStar, which filled the gap in his global satellite coverage.

It was not the same, however, as deploying Deathstar.

"I was very depressed by it," said Murdoch, "but it was a matter of discretion over valor."

A fitting epitaph.

———

HAD IT SURVIVED, Deathstar could have been a commercial bloodbath for all the companies involved—all to the benefit of consumers. As $50 Sky satellite dishes popped up on rooftops and windowsills, with the possibility of local broadcast channels, cable companies would have had no choice but to cut prices, add channels, improve service and fight for business.

As it was, cable rates were deregulated on March 31, 1999.

It was the cable gang's move. The safe was cracked. The door was wide open. All they had to do was raise rates slowly and take the money out a little at a time so as not to wake the snoozing guard— that is, Congress, which had set the deregulation date in the 1996 Telecommunications Act and was loath to change it.

Because satellite TV and other competitors had whittled cable's share of the subscription television market down to 85 percent, it was said that effective competition existed, even though Congress had yet to resolve the legality of satellite TV delivering local broadcast channels.

The cable buzzword was "restraint." Rate hikes announced for 1999 averaged 5 percent, higher than the overall inflation rate but lower than the previous three years, which totaled 25 percent. Yet, cable operators large and small were free to charge what they liked, to test the vagaries of supply and demand. At what monthly price would subscribers begin dropping a pay channel, or drop cable completely, or buy a satellite dish?

Would cable companies focus on new moneymaking services? Would they go slow on cable rate hikes? Or would it be 1987 all over again, the last time rate controls were lifted and some operators got greedy and the safe door was slammed by Congress, inaugurating a decade-long mishmash of regulation?

Whatever the result, the composition of the cable gang had irrevocably changed, coalescing into bigger and bigger conglomerates, just as the Las Vegas casino industry did in the 1970s when the likes of Hilton and Holiday Inn began taking over from the colorful band of cutthroats who built the neon palaces in the desert.

In cable, AT&T, Microsoft and Paul Allen were the outside change agents.

And then there was this tangle: In March of 1999, Comcast launched a $58 billion deal for MediaOne, the Colorado cable company with five million subscribers, made up largely of the former Continental Cablevision of Boston. Key shareholder Amos Hostetter, Continental's founder, objected to the terms of the Comcast offer and sought out other bidders—which drew in AT&T, still swallowing TCI, but hungry enough to make a counteroffer for MediaOne of $62 billion. That caused Comcast to seek out Microsoft and even America Online, but that alliance didn't work. AT&T bought MediaOne and a cable company called Lenfest, then sold a two-million-subscriber cluster to Comcast, which also received a MediaOne breakup fee of $1.5 billion. Then Microsoft invested $5 billion in AT&T to get further enmeshed in its digital set-top box.

But that is just one scorecard from the interlocked game, which is going on right now, all over the world.

———

JOHN MALONE, WHO had studied thermodynamics in school, was fond of satellite launch analogies. When Liberty Media's stock took off in 1991, he described it as rising like a Saturn 5. When TCI came out of its tailspin in early 1997, Malone said, "When they launch a satellite into space, they adjust the trajectory all the way. They don't just shoot the rocket off and hope it goes where it's supposed to go. Well, companies are the same way. That's the answer I give to people who say, what the hell's going on at TCI? We're making minor midcourse corrections."

Malone's three decades in the cable business had the aspect of a satellite launch. There was the shower of fire, the slow liftoff, the steady ascent and then the disappearing act when the satellite itself locks into the Earth's orbit, out of sight, sending and receiving signals as needed.

The chess machine's capsule was Liberty Media, which he kept as TCI's cable operations, @Home and other assets were handed over to AT&T. Malone not only became AT&T's largest individual shareholder and a board member, he struck a phenomenal deal to maintain control of Liberty. AT&T gave him $5.5 billion to play with, plus the ability to borrow $12 billion more. Given Malone's penchant for debt leverage, he could use the money to buy more and more pieces of more and more programming, technology and Internet companies. Liberty also had exclusive access to one analog channel on all of AT&T's cable systems, which could translate to 10 or more digital channels in the future. So, while Malone receded from public view, his dealings will affect a phone line, a cable service and an Internet connection near you.

"It's been a hell of a run and now we're going off to a new adventure," he said on February 17, 1999, the day TCI shareholders approved the merger with AT&T during a meeting at TCI's National Digital Television Center.

TCI's share price that month was $64. Two years earlier, with TCI back on its heels and Deathstar bearing down, the price was under $13. Liberty Media shares traded at $56 in early 1999, four times the price of the darkest days in 1996.

Malone's net worth experienced a similar multiplier effect as he exited the cable industry, rising to an estimated $4 billion, primarily in AT&T and Liberty stock.

After the shareholder meeting was over, Malone walked to a press conference with Dob Bennett and Gary Howard, two top Liberty officials who had backgrounds in finance and would be implementing the deals. Malone batted press questions over the fence and reeled off some accumulated wisdom: "You've got to be hedged in all directions. Bet on all sides of all fences. You want to be an industry leader or create one."

His dream had come true. He had everything he wanted. He ran a portfolio company that was purely about moving assets and building wealth. The chess machine was in the glorious orbit of friction-free capitalism.

—

CHARLIE ERGEN SHIFTED into a padded, black swivel chair in the EchoStar boardroom and laughed at the reporter's tape recorder on the long conference table.

"Is this a deposition?" he asked. "I take the Fifth. I don't remember."

It was two days before New Year's Day, 1999. Ergen was preparing to leave for Arizona to visit the Sky satellite uplink that Murdoch built and Ergen will own. Then to the Fiesta Bowl where Ergen would watch his alma mater, the University of Tennessee, win the national collegiate football championship.

He was dressed casually, in jeans, a patterned sweater and sneakers. At a $1,000-a-plate fundraiser for Senator John McCain several months earlier in downtown Denver, Ergen wore a similar outfit, but with a shirt emblazoned with the Dish Network logo. The host of the event was Philip Anschutz, the billionaire founder of Qwest Communications.

"These are the leaders of the Denver community, so I'm going to be selling Dish," said Ergen, voice rising as he defended his attire. "I sold that day! I sold Anschutz one at lunch."

Ergen lived to sell. It wouldn't matter, and almost didn't, whether it was satellite dishes or eelskin wallets. His company's stock was due to blast off, from $25 in late 1998 to beyond $150 in the second quarter of 1999 before a stock split. Ergen's net worth was at $3 billion and climbing.

With a few reversals, Ergen played his cards just right. During a boom economy, he built his entire company around the digital future and the satellite pipes to deliver it. EchoStar had two million subscribers and was adding 100,000 a month. With two more satellites due to launch in 1999—both to the embattled 110 slot—EchoStar would have six birds in the sky, with enough capacity for 500 channels and a range of data services to the home. Ergen could align with the Baby Bells against the AT&T cable/phone/Internet juggernaut, remain a free agent or someday sell his company to the highest bidder.

Sitting in the boardroom, he explained the way the game played out:

> Malone got what he wanted. He sold to a phone company and he
> gets $5.5 billion to go run a content company. He likes us now.

We're an ally now. Murdoch got what he wanted. He got The Family Channel. He got Fox Sports going. He got Fox News going. He got out of a situation that wasn't going to be a good situation in the U.S. He's also a content guy who likes us and likes cable. Murdoch can say, "I'm a neutral guy." Malone can say, "I'm a neutral guy." They both got out of the distribution business, which they see as riskier and less upside than programming. They're right about that. We're in the distribution business, which is what we've always done. So, it's been a good game.

Like poker?

"It's very difficult to play poker with someone the first time," he explained. "It takes a lot of discipline. You can't play on emotion. It takes a lot of reading the competition, your opponents. It gets easier as you play with them." He paused. "I never bluff. Well, almost never. I've been the sucker in a few games, but the rules are very similar rules to business. You bet big when you've got a big hand."

And who was Charlie Ergen? Was he using satellite technology to liberate the masses from the tyranny of cable? Was he a little guy fighting the system? Or was it an effective pose, the ultimate leverage by a country boy from Tennessee who used the government's antagonism toward the cable gang to get the better of two players of superior force, Murdoch and Malone?

Who was Elvis playing for all along?

Himself, of course.

That's how he became a cutthroat.

NOTES

C itations below from *The Denver Post* between 1996 and 1999 without a reporter's name attached are stories written for the newspaper by the author. In addition, all interviews and notes are the author's work unless otherwise indicated, with all material compiled from 1996 to 1999. In some instances below—such as citations from *Broadcasting* magazine through the early 1990s—the original article did not carry a byline, or the reporter's name was unavailable from archives. For more details, or to contact the author, please send e-mail to stephen@keating.org.

Front matter epigraph from "Antitrust," copyright © 1962 by Alan Greenspan, from *Capitalism: The Unknown Ideal*, by Ayn Rand. Used by permission of Dutton Signet, a division of Penguin Putnam Inc.

Prologue: The Game

The word "cutthroat," which can mean murderous or ruthless, also refers to a type of game played by individuals rather than by partners, as in bridge and racquetball. The tactics employed by corporate chiefs owe a debt to military strategists dating back to Sun Tzu, circa 500 B.C. The Chinese general wrote that "all warfare is based on deception," and "the supreme art of war is to subdue the enemy without fighting."

Analysis of interlocked media companies is prevalent in the popular and academic press. For one source, see *The Media Monopoly*, fifth edition, by Ben Bagdikian (Beacon Press, 1997).

The three books listed below were particularly helpful in framing the present subject. *The Highwaymen: Warriors of the Information Superhighway* by Ken Auletta (Random House, 1997), a collection of his writings for *The New Yorker* magazine between 1993 and 1996. Three of his terms—

the pirate, the consigliere and the magic box—were adapted from that book, as well as specific citations in the text and below. *News and the Culture of Lying: How Journalism Really Works* by Paul H. Weaver (Free Press, 1994), an examination of the news media. *Moral Calculations: Game Theory, Logic, and Human Frailty* by László Mérő (Copernicus, 1998), which explains the psychology and mathematics of risk-taking.

Chapter 1: The Cable Gang Mourns

The author attended the funeral with other reporters and co-wrote an article on it with Joanne Davidson of *The Denver Post*, published 11/21/96.

p. 2 "I'd have to think harder ...": Daniel A. Beucke, *The Denver Post*, 4/1/84.

p. 3 Ten men on the *Forbes* list ... : *Forbes*, 10/13/97.

p. 3 DirecTV became the fastest-selling ... : Consumer Electronics Manufacturing Association.

p. 5 "TCI is purely and simply ...": *The Denver Post*, 4/3/96.

p. 5 "By directing that industry ...": Adam Smith, *An Inquiry into the Nature and Causes of the Wealth of Nations* (Random House edition, 1937), p. 423.

p. 5 "I would call myself ...": Joanne Ostrow, *The Denver Post*, 6/6/93.

p. 6 "I've probably given ...": All quotes from the funeral are from the author's notes.

p. 7 Al Gore, as a rising U.S. senator ... : transcript of Subcommittee on Communications, U.S. Senate Committee on Commerce, Science and Transportation, 11/17/89, p. 407.

p. 8 "I've got him right where ...": Jolie Solomon, *Newsweek*, 10/25/93.

p. 8 "I'm being clitorized.": transcript of Ted Turner's remarks to the National Press Club, 9/27/94, reported among other places, *Electronic Media*, 10/10/94.

p. 9 His greatest coup was ... : *The Denver Post*, 11/17/96.

Chapter 2: Deathstar

Description of the 2/24/97 event at the Fox lot is taken from author interviews of participants and review of a videotape of the event, courtesy of *Satellite Business News*.

p. 14 After graduating from Oxford ... : *The Guardian* newspaper, 11/9/98.

p. 14 Murdoch viewed the 1974 resignation ... and the following anecdotes: Thomas Kiernan, *Citizen Murdoch* (Dodd, Mead & Company, 1986), pp. 144, 232–236, 260.

p. 14 A News Corp. subsidiary contributed ... : covered, among other places, by Ken Auletta, *The Highwaymen*, pp. 282–285.

p. 15 The latter innovation ... and the following quotes: Kiernan, *Citizen Murdoch*, p. 126.

p. 15 Early features included ... : ibid., p. 153.

p. 16 "He's like a great white shark ...": *The Denver Post*, 4/3/96.

p. 16 "We're taking a friendly approach.": John Lippman, *Los Angeles Times*, 2/28/92.

p. 16 "the ultimate example ...": Richard Zoglin, *Time*, 6/27/94.

p. 16 "If the industry wants to spend ...": Bill Carter, *The New York Times*, 6/6/93.

p. 17 "I would have liked to start ...": Joe Flint, *Broadcasting*, 1/17/94.

p. 18 "Right now Rupert Murdoch gets ...": Gary Samuels, *Forbes*, 12/19/94.

p. 18 "Fox News is going to be ...": *The Denver Post*, 10/7/96.

p. 19 "shakedown" by TCI ... : transcript of Subcommittee on Communications, U.S. Senate Committee on Commerce, Science and Transportation, 11/17/89, p. 407.

p. 19 Yet, a $100,000 donor ... : Bob Woodward, *The Washington Post*, 3/2/97.

p. 20 "If these guys want to go ...": Bryan Burrough and Kim Masters, *Vanity Fair*, 1/97.

p. 20 "HEY ... TED ...": ibid.

p. 20 "Rupert's worth $3.5 billion. ...": *The Denver Post*, 11/15/96.

p. 21 "I think we're moving ...": *The Denver Post*, 1/26/96.

p. 22 Boesky actually said ... : James Stewart, *Den of Thieves* (Touchstone, 1992), p. 261.

p. 22 "This deal will change ...": interview with Charlie Ergen.

p. 24 On February 7, 1997 ... : interview with Judianne Atencio.

p. 24 "Those cable guys are greedy ..." and "Chahlee ...": interview with Ergen. Rupert Murdoch declined to be interviewed.

p. 24 "He has this fatal capacity ...": William Shawcross, *Murdoch* (Simon & Schuster, 1992), p. 174.

p. 25 "You know what …" and details of Murdoch visit to Ergen home: interview with Candy Ergen.

p. 25 "He can't fire me …": Mark Robichaux, *The Wall Street Journal,* 2/26/97.

p. 26 "We're aiming for the big …" and next two quotes: from author's notes of Murdoch's press call on 2/24/97.

p. 27 "He's crazy," Malone told … : interviews with TCI officials.

p. 27 "Tell 'em I'm …": interviews with EchoStar officials.

p. 27 "The problem is …": Jim Carrier, *The Denver Post,* 9/22/96.

p. 28 … Malone, who claimed that … : transcript of Subcommittee on Communications, U.S. Senate Committee on Commerce, Science and Transportation, 6/14/89, p. 14.

p. 28 "Given the monopoly character …": transcript of Subcommittee on Communications, U.S. Senate Committee on Commerce, Science and Transportation, 11/17/89, p. 302.

p. 28 "The easiest way …" and subsequent Padden quotes: transcript of his remarks on 2/24/97.

p. 30 "The presentation of the deal …" and Murdoch's reaction: interview with Gordon Crawford.

p. 30 Yet, Crawford was one of the doubters … : Jerome Tuccille, *Rupert Murdoch* (Donald I. Fine, 1989), p. 134.

p. 31 "Rain of Terror?": Tom Kerver, *Cablevision,* 4/7/97.

p. 31 "We're going to make it as tough …" and the following quote from Glenn Jones: Fred Dawson and John Higgins, *Multichannel News,* 3/24/97.

p. 31 Marcus Cable in Dallas refused … and "If someone is threatening …": Sallie Hofmeister, *Los Angeles Times,* 3/12/97.

p. 32 "I remember having conversations …": interview with Trygve Myhren.

p. 32 "I doubt that they will get along …": Bob Scherman, *Satellite Business News,* 3/24/97.

Chapter 3: Stealing Free TV

The chapter title comes from Paul Maxwell of Media Business Corp. in Golden, Colorado. The story of the creation of the Worland cable system comes from separate author interviews with Roy Bliss Sr. and Tom Mitchell, supplemented by a transcript of an interview with them conducted in 1991 by Robert Allen for the National Cable Television Center and Museum. Cable Center interviews cited below were used with permission.

p. 38 "In one sense, the local operator ...": interview with Stratford Smith.

p. 39 "The theater, the world of good music ...": 4/8/69 letter to Dillon Cable TV.

p. 39 One of the nation's first cable ... : George Mannes, *American Heritage of Invention & Technology*, reprinted in *Multichannel News* on 12/2/96.

p. 39 In Mahanoy City, Pennsylvania ... : Thomas P. Southwick, *Distant Signals: How Cable TV Changed the World of Telecommunications* (Primedia Intertec, 1999), p. 12.

p. 39 Viewers tuned into shows ... : listing of TV shows, *The New York Times*, 12/22/50.

p. 39 It was that year when Bill Daniels ... : Recollections and quotes of Daniels's early years are taken primarily from a transcript of an interview with him conducted in 1986 by Max Paglin for the cable center, pp. 10, 18–19.

p. 40 "We sent out ballots ...": interview with Gene Schneider.

p. 40 The staccato paragraph ... : *Forbes*, 10/18/93; Copyright 1993 *Forbes* magazine. Reprinted with permission.

p. 41 "Take a look at where ...": author's notes of 11/20/97 event at Cableland.

p. 41 "We were just scratching. ...": interview with Daniels.

p. 41 Magness was born ... : Recollections and quotes of Magness's early years are taken primarily from a transcript of an interview with him conducted in 1989 by Stratford Smith for the cable center, pp. 1–16.

p. 42 "If a system is operated ...": *TV Digest*, 10/3/53.

p. 43 "What we have here ...": *Broadcasting*, 6/25/62.

p. 43 "The broadcasters wanted it ...": interview with Schneider.

p. 44 "I have often said ...": Paglin interview with Daniels, p. 28.

p. 44 Kahn became an unlikely ... : Irving Berlin Kahn's background taken primarily from an article by Peter W. Bernstein, *Fortune*, 7/28/80.

p. 45 they exhorted each other ... and "America does not need ...": *Broadcasting*, 4/13/64.

p. 45 Early on, Weaver challenged ... : ibid.

p. 46 An impertinent questioner ... : *Broadcasting*, 5/11/64.

p. 46 Teleglobe tested ... : *Broadcasting*, 10/8/62.

p. 46 Home Box Office forever divided ... and details of HBO's early years: taken primarily from "HBO: The First 25 Years: An Anniversary Salute," written by Kathy Haley and published in 1997 by *Cablevision* and *Multichannel News.*

p. 47 used Comsat's *Early Bird* ... : Anthony Michael Tedeschi, *Live via Satellite: The Story of COMSAT and the Technology That Changed World Communication* (Acropolis Books, Ltd., 1989), pp. 40–41.

p. 48 From sleepy Atlanta ... .and details of Ted Turner's early years: Porter Bibb, *Ted Turner: It Ain't as Easy as It Looks* (Johnson Books, 1997), pp. 102–107.

p. 48 One day in 1976 ... : The story of Taylor Howard's home satellite TV system comes from author interviews with Howard.

p. 50 On the cover of its 1979 Christmas catalog ... : text of Neiman Marcus ad reprinted in "Direct-To-Home Satellite Broadcasting," a 1980 compilation of articles by John P. Taylor, from *Television/Radio Age.*

p. 51 A 1979 *Forbes* article: Allan Sloan, *Forbes*, 12/10/79.

p. 51 One backhaul showed former CIA ... : Jefferson Graham, *USA Today*, 2/25/92.

p. 51 Another showed Max Robinson ... : recollection of Taylor Howard. Max Robinson died in 1988.

p. 51 Comedians including Johnny Carson ... : Graham, *USA Today*, 2/25/92.

p. 52 "Virtually everybody but TCI ...": interview with Stratford Smith.

p. 52 While patronage and political fixes ... : A primary source on the franchise battles was "The Gold Rush of 1980," a lengthy article in *Broadcasting*, 3/31/80.

p. 53 "In 1983, along with Cablevision ...": Bill Frezza, *Internet Week*, 1/18/99; used with permission.

p. 53 "an orgy of excesses.": John Cooney, *Fortune*, 4/18/83.

p. 53 "When politics are involved ...": Dennis Schatzman, editor of *The New Pittsburgh Courier*, in *Business Week*, 2/18/80.

p. 54 The five winning bidders who divvied up ... : "The Gold Rush of 1980," *Broadcasting*, 3/31/80.

p. 54 "If ever there was an argument ..." and The Cincinnati school board ... : ibid.

p. 54 In Denver, 22 individuals ... and "The best offer ...": Maurine Christopher, *Advertising Age*, 6/7/82 and 5/3/82.

p. 55 "I really started to worry ...": interview with Trygve Myhren.

p. 55 "helter-skelter": *Broadcasting*, 3/31/80.

p. 55 "a consumer will have ...": transcript of Subcommittee on Communications, U.S. Senate Committee on Commerce, Science and Transportation, 2/17/89, p. 303, referring to Thomas Wheeler testimony from 2/83.

p. 55 "The municipal franchise is issued ...": Thomas Hazlett, *The New York Times*, 11/17/85.

p. 55 Warner Amex won Dallas ... and "We felt a new name ...": Todd Mason, *Business Week*, 7/22/85.

p. 55 "We weren't going to build libraries ...": Bill Powell and Janet Huck, *Newsweek*, 6/1/87.

p. 56 Magness said later that corruption ... : Magness interview by Smith, p. 129.

p. 56 One call for draining the ... : *Broadcasting*, 9/22/80.

Chapter 4: The Chess Machine

The story of John Malone's early years is drawn from research of records in Milford, Connecticut, the Hopkins School in New Haven and Yale University in New Haven, supplemented by interviews and other sources where indicated.

p. 59 John Malone's family had the first ... : interview with Edward Kozlowski. Also reported by Frank Juliano, the *Connecticut Post*, 6/27/98.

p. 59 He later tinkered ... : Mark Ivey, *Business Week*, 10/26/87.

p. 59 "an intellectual with white socks.": Ken Auletta, *The Highwaymen*, p. 29.

p. 59 "Whatever was on her mind ...": interview with Kozlowski.

p. 60 "Nothing distinctive. ..." and Yale as "socialistic": Auletta, *The Highwaymen*, pp. 29–30.

p. 60 Malone ran track ... : interview with Fred Andreae.

p. 60 They had been introduced ... : Auletta, *The Highwaymen*, p. 30.

p. 60 "My background is German ...": author's notes of Malone remarks to a Cato Institute luncheon, 11/15/96.

p. 61 "John thought through ...": interview with Mandell Bellmore.

p. 61 "Given that it was a time ...": interview with George Nemhauser.

p. 62 "Son, this is Talmudic wisdom. ...": Edward W. Desmond, *Fortune*, 2/16/98.

p. 62 Shapiro passed up Malone … : interview with John Sie, a TCI executive and former Jerrold engineer.

p. 62 "They treated me …": author's notes of Malone's remarks at Bob Magness's funeral.

p. 63 "We are really attempting …": Smith interview with Magness.

p. 63 Yet, one particular cable system … : interview with J. C. Sparkman, a TCI executive.

p. 63 He helped secure … : Jack Phinney, *The Denver Post,* 6/1/73.

p. 63 "Vail was born …" and Vail's early history: June Simonton, *Vail: Story of a Colorado Mountain Valley* (Taylor Publishing Co., 1987), pp. 10, 58, 91–97; used with permission.

p. 64 "If we lost this …": Peter W. Bernstein, *Fortune,* 7/28/80.

p. 64 One Thursday night … : The 1973 incident in Vail is reconstructed primarily from an interview with Terrell Minger, and reports in *The Denver Post,* 11/5/73 and 11/7/73.

p. 65 "We were demonstrating …" and the reference to TCI's internal memo and "I didn't want the reputation …": deposition of John Malone conducted on 8/20/84 and 8/21/84 in *Central Telecommunications Inc. vs. TCI Cablevision Inc., et. al.,* civil case 83-4068-CV-W-5, U.S. District Court for the Western District of Missouri, Central Division.

p. 65 "There was a reluctance …": interview with Smith.

p. 65 One night in the mid-1970s … and "My face looked like …": interview with J. C. Sparkman.

p. 66 They borrowed money … : John Higgins, *Multichannel News,* 4/1/91.

p. 66 One day in 1976 … and subsequent quotes by Malone: interview with Malone.

p. 67 "We did almost all …": interview with Lorraine Spurge.

p. 67 In 1979, TCI shareholders agreed … : Jack Phinney, *The Denver Post,* 7/20/79.

p. 67 A cable analyst named Jack Myers … and "In the first 10 minutes …": Martha T. Moore, *USA Today,* 10/14/93.

p. 68 "Bob seemed like the kind of guy …": Derek T. Dingle, *Black Enterprise Titans of the B.E. 100s: Black CEOs Who Redefined and Conquered American Business* (John Wiley & Sons, 1999), p. 38.

p. 69 "When you really start showing earnings …": interview with Malone.

p. 70 "What I really love is strategy": John Higgins and Price Colman, *Broadcasting & Cable*, 7/13/98.

p. 70 Disney chief Michael Eisner once recalled ... : Michael Eisner, *Work in Progress* (Random House, 1998), p. 7.

p. 71 "To call us a giant ...": Joanne Ostrow interview notes with Malone, some of which appeared in *The Denver Post*, 6/6/93.

p. 71 "The garbage wouldn't be on ...": author's notes of Malone remarks to a Cato Institute luncheon, 11/15/96.

p. 71 "We just kind of laughed ...": interview with Frank Biondi.

p. 71 "documentary freak" and "great comedy": *The Denver Post*, 4/3/96.

p. 71 "the most valuable media property ...": David Lieberman, *USA Today*, 2/12/97.

p. 72 "Little kids using foul ...": author's notes of Malone remarks at a TCI event, 3/19/98.

p. 72 "a social libertarian ...": Cato Institute luncheon, 11/15/96.

p. 73 "We refuse to get raped ...": Christopher Knowlton, *Fortune*, 7/31/89.

p. 73 "Citizens in a community don't care ...": *Broadcasting*, 3/31/80.

p. 73 TCI, however, would face a costly ... : Details of the Jefferson City, Missouri, case are taken from court records and interviews with Elmer Smalling and Stephen Long.

p. 74 The company was so broke ... : Jim Cooper, *Cablevision*, 9/4/95.

p. 75 Then came the Turner bailout. ... : Porter Bibb, *Ted Turner: It Ain't as Easy as It Looks* (Johnson Books, 1997), pp. 293–297, 316–318.

p. 75 "Ed, you're the seventy-fifth ...": interview with Ed Taylor.

p. 75 A *Fortune* story ... : Christopher Knowlton, *Fortune*, 7/31/89.

p. 75 A 1992 story ... : Johnnie L. Roberts, *The Wall Street Journal*, 1/27/92.

p. 76 his wife Betsy died ... : *The Denver Post*, 9/24/85.

p. 76 a woman named Sharon Costello ... : The account of Costello's background and relationship with Bob Magness comes from a one-page biography supplied by her.

p. 76 They began living together ... and financial details: pre-nuptial agreement filed as part of the Magness estate civil case, 96PR944, District Court, Arapahoe County, State of Colorado.

p. 76 He owned a 59-foot sailboat ... : Mark Ivey, *Business Week*, 10/26/87.

p. 76 In the late 1990s … : Erin Emery, *The Denver Post*, 4/26/98.

p. 77 In 1990, TCI invested … : Alec Foege, *The Empire God Built* (John Wiley & Sons, 1996), pp. 43–46.

p. 77 Sie was a native of Shanghai … : interview with John Sie.

p. 78 This so-called negative option … : Kevin Goldman, *The Wall Street Journal*, 5/29/91.

p. 78 "It was a good idea …": interview with Sie.

p. 78 "There's nothing wrong …": interview with Malone.

p. 78 Rockefeller did it on a much larger scale … and subsequent Rockefeller examples: Ron Chernow, *Vanity Fair*, 5/98. See also Chernow's biography, *Titan: The Life of John D. Rockefeller Sr.* (Random House, 1998).

p. 79 "Rockefeller envisioned a new …": Chernow, *Vanity Fair*, 5/98.

p. 79 Standard Oil was broken up … and … Ida Tarbell wrote … : ibid.

p. 79 "Thus, on one hand …": Ida M. Tarbell, *The History of the Standard Oil Company* (McClure, Phillips and Co., 1904), pp. 102–103.

Chapter 5: Cutting Cable

The early days of Echosphere were reconstructed primarily from separate interviews with the company co-founders: Charlie Ergen, Candy Ergen and Jim DeFranco.

p. 82 Oak Ridge was built and run … : Charles W. Johnson and Charles O. Jackson, *City Behind a Fence: Oak Ridge, Tennessee, 1942–46* (University of Tennessee Press, 1981), pp. 3–12, 71.

p. 82 Charlie Ergen's family history and youth in Oak Ridge: interviews with Ergen, supplemented by an interview with Dick Smyser, founding editor of *The Oak Ridger* newspaper.

p. 83 On the night of October 4, 1957 … : *The Denver Post*, 10/5/97.

p. 88 "The main thing he wanted …": interview with Bob Luly.

p. 89 "TCI was so poor …": author's notes of Malone remarks at a TCI event, 3/19/98.

p. 89 The son of a famed U.S. senator … and Gore's early background: Hank Hillin, *Al Gore Jr.: His Life and Career* (Birch Lane Press, 1992), pp. 13–26, 52–59, 70–71, 97–101.

p. 90 His travels took him to The Farm … : interview with Mark Long.

p. 91 "He was excited when he got a watch …" and subsequent quotes from Mike Kopp: interview with Kopp.

p. 91 In the 1984 Cable Act ... and subsequent references to Gore's advocacy on related issues: primarily from "The Role of Senator Albert Gore Jr. in Satellite/Cable Legislation," a paper presented in August of 1990 by Michael B. Doyle of Arkansas State University, Jonesboro, to the Communication Technology and Policy Group, AEJMC annual convention, in Minneapolis, Minnesota.

p. 92 "This bill removes the legal cloud ...": Carolyn Shoulders, *The Tennessean*, 10/2/84.

p. 92 Porn channels on satellite ... and subsequent quotes from Dawson and Smith: *Broadcasting*, 8/6/84.

p. 92 "We suffer to some degree ...": *Broadcasting*, 2/24/86.

p. 93 Quotes from Rose and Nielson ... : *Broadcasting*, 3/10/86.

p. 93 In the early morning hours ... and HBO hacking incident: Anne Wells Branscomb, *Who Owns Information?* (Basic Books, 1994), pp. 106–111.

p. 94 "He was preaching the gospel ...": transcript of Subcommittee on Communications, U.S. Senate Committee on Commerce, Science and Transportation, 11/17/89, p. 407.

p. 95 "We couldn't afford ...": interview with Tim Robertson.

p. 95 "That's bullshit. ...": interview with Roy Neel.

p. 95 In fact, the pressure on Turner ... : interview with John Sie.

p. 95 "The marketplace for home satellite ..." and "We were too ..." and subsequent quotes from hearing: transcript of Subcommittee on Communications, U.S. Senate Committee on Commerce, Science and Transportation, 7/31/87, pp. 4, 154–156, 194.

p. 96 Eight months after the hearing ... and subsequent details: U.S. Customs reports, interviews with EchoStar officials and *Satellite Business News*, 6/13/90, 12/12/90 and 4/3/91.

p. 97 "We weren't buttoned up ...": interview with Ergen.

p. 98 "Where's your passion?": recounted by Kopp.

p. 98 "We never operated ...": interview with Kopp.

p. 98 MultiVision raised rates ... and "It's good news ...": *The Tennessean*, 6/21/89.

p. 98 "The cable industry has been praying ...": *Daniels Letter*, a cable industry newsletter published by Bill Daniels, 4/86.

p. 99 "Some operator is going to ...": *Daniels Letter*, 12/86.

p. 99 "I would want the whole country ...": Shirley Nanney, *The Nashville Banner*, 8/23/89.

p. 99 "Meet the man who makes ...": Mark Ivey, *Business Week*, 10/26/87.

p. 99 "cold, business-like ...": interview with Neel.

p. 100 "I knew we had lost the battle ...": interview with Sie.

p. 100 "To convey to a handful ...": transcript of Subcommittee on Communications, U.S. Senate Committee on Commerce, Science and Transportation, 5/11/88, pp. 397–398.

p. 101 "Some just say, 'Well, that's the marketplace ...'" and subsequent dialogue between Gore and Malone: transcript of Subcommittee on Communications, U.S. Senate Committee on Commerce, Science and Transportation, 11/16/89, pp. 4, 108–110, 161–162.

p. 103 "trying to get Samoan warlords ...": interview with Glenn Jones.

p. 103 "John has a favorite saying ...": Laura Landro, *The Wall Street Journal*, 1/22/90.

p. 105 "Cable consumers are getting ripped off ...": Edmund L. Andrews, *The New York Times*, 10/4/92.

p. 105 "Four more weeks! ...": recollection of Bob Scherman, *Satellite Business News*.

p. 105 "Bush's veto was anti-consumer ...": interview with Neel.

p. 106 rate regulation was a gimmick ... : For an extended discussion on cable rate regulation, see *Public Policy Toward Cable Television: The Economics of Rate Controls* by Thomas W. Hazlett and Matthew L. Spitzer (The MIT Press, 1997).

p. 106 "We'll continue to diversify away ...": David Kline, *Wired*, 7/94.

Chapter 6: The Rat Zapper

Arthur C. Clarke's early background: Neil McAleer, *Arthur C. Clarke: The Authorized Biography* (Contemporary Books, 1992), pp. 1–15, 25–35, 53–57.

p. 108 Clarke envisioned manned space stations ... and quotes: from Clarke's initial manuscript, 5/45, edited for *Wireless World*, 10/45.

p. 109 "I suspect that my early ...": Clarke's acceptance speech on receiving the Marconi Award in 1982.

p. 109 Comsat's funding came from ... and early company history: Anthony Michael Tedeschi, *Live via Satellite*, pp. 25–35.

p. 111 "Direct by satellite is never going to be competitive ...": *Cablevision*, 7/81.

p. 111 Malone referred to DBS ... : interview with Eddy Hartenstein, DirecTV president.

p. 111 required international agreements ... : *Broadcasting*, 7/25/83.

p. 112 "Retailing and TV programming ...": *Business Week*, 7/19/82.

p. 113 "If nobody wants to subscribe ...": *Broadcasting*, 9/15/80.

p. 113 "corporate Vietnam": Sarah Bartlett, *Fortune*, 4/18/83.

p. 113 Johnson was a U.S. Navy veteran ... and the following quotes: interview with Robert W. Johnson.

p. 113 "In light of its potentially disruptive ...": *Broadcasting*, 7/14/80.

p. 114 USCI was backed ... and Hickey quote: Bartlett, *Fortune*, 4/18/83.

p. 115 "It had very limited ...": interview with John Malone.

p. 115 TCI would turn USCI ... : interview with John Sie.

p. 115 But Crimson bled red ... : *Broadcasting*, 12/12/88.

p. 116 "One reason they wanted Tempo ...": interview with Sel Kremer.

p. 116 When Malone saw that ... : interview with Ed Taylor.

p. 116 "TCI has a demonstrable track record ...": *Broadcasting*, 6/27/88.

p. 116 "There are really no limitations ...": Thomas Kiernan, *Citizen Murdoch*, p. 272.

p. 117 "I said, 'Rupert ...'": interview with Dan Ritchie.

p. 117 Details of Skyband's cost and "Do you know anybody ...": *Broadcasting*, 11/14/83.

p. 117 "If they think they can beat me ...": Kiernan, *Citizen Murdoch*, p. 284.

p. 117 It grew out of frenzied research on High Definition Television ... and digital compression background: Joel Brinkley, *Defining Vision: The Battle for the Future of Television* (Harcourt Brace, 1997), pp. 120–125.

p. 118 At a February 1990 press conference ... and subsequent quotes and "If the Sky Cable partners wanted to downplay ...": *Broadcasting*, 2/26/90.

p. 119 By the end of 1990 ... and News Corp.'s financial struggles: William Shawcross, *Murdoch*, pp. 13–26.

p. 120 Tom Rogers, an NBC executive ... : *Broadcasting*, 2/11/91.

p. 120 On the day Sky Cable ... : *Broadcasting*, 2/26/90.

p. 120 Primestar was a "transitional ...": *Broadcasting*, 1/29/90.

p. 120 "I had a letter of intent ...": interview with Charlie Ergen.

Chapter 7: 500 Channels

p. 121 "One of the biggest problems ...": Stephen Keating, *Cablevision*, 2/22/99.

p. 122 "The only reason ..." and subsequent quotes from John Cusick: interviews with Cusick.

p. 122 The backers of SkyPix ... : Graham Button, *Forbes*, 7/20/92.

p. 122 At the National Cable Show ... and "highly realistic ...": John Higgins, *Multichannel News*, 4/1/91.

p. 123 "Television will never be the same." ... : Edmund L. Andrews, *The New York Times*, 12/3/92.

p. 124 His 1993 appearance ... : transcript of the *TechnoPolitics* show, 4/93.

p. 124 "Five hundred channels, movies on demand ...": Kevin Maney, *USA Today*, 3/30/93; copyright 1993, *USA Today*. Reprinted with permission.

p. 124 "It was new and exciting. ...": interview with Maney.

p. 125 "Malone is famous ...": interview with Jim Chiddix.

p. 125 Asked about the legacy ... : *The Denver Post*, 12/15/96.

p. 126 "Why the hell ...": interview with J. C. Sparkman. Robert Stempel declined comment.

p. 126 GM announced it was ... : Doron P. Levin, *The New York Times*, 12/4/92.

p. 126 "I'll never forget it ...": Keating, *Cablevision*, 2/22/99.

p. 127 He once called Primestar ... and subsequent quotes from Eddy Hartenstein: interviews with Hartenstein.

p. 127 a family of firsts. ... : USSB company history.

p. 128 "The message that I'm giving ..." and "We don't want to be riding ...": *Broadcasting*, 7/20/87.

p. 128 "My father saw DBS ...": interview with Stanley E. Hubbard.

p. 129 Before the launch, a Comcast cable executive ... : recollection of John Higgins, *Broadcasting & Cable*.

p. 129 In June of 1993 ... : Justice Department consent decree, civil case 93-CIV-3913, U.S. District Court, Southern District of New York.

p. 129 "A lot of theories ...": interview with James Rill.

p. 131 Redstone had been ... : *Forbes* 400 richest Americans, 10/19/92.

p. 132 Viacom filed a 91-page lawsuit ... and subsequent quotes: from the lawsuit.

p. 133 "In a world where ...": Joanne Ostrow interview notes with John Malone, some of which appeared in *The Denver Post*, 6/6/93.

p. 134 "He talks at 60 miles ...": *The Denver Post*, 2/14/99.

p. 134 "You need to write ...": conversation with the author.

p. 134 "I've got my uniform ...": Dinah Zeiger, *The Denver Post*, 2/25/94.

p. 134 "The Plot to Cripple Cable" ... : Tom Kerver, *Cablevision*, 7/3/95.

p. 136 "Time Warner was terrible ..." and "This has to be ...": Keating, *Cablevision*, 2/22/99.

p. 136 "We have an ownership interest ...": Don West and Harry Jessell, *Broadcasting & Cable*, 11/26/94.

p. 136 "I don't want to discourage ...": Don West and Joe Flint, *Broadcasting & Cable*, 1/24/94.

p. 137 "Every time we get units in ...": *Satellite Week*, 7/18/94.

Chapter 8: Fighting for Air

Quotations and background of Dan Garner come from interviews with Garner and associates.

p. 140 A company called Home Broadcasting ... : Robert N. Wold, *Via Satellite*, 9/96.

p. 141 "Dan had a license ...": interview with Gordon Apple.

p. 141 "I considered making ...": interview with Jackson T. Stephens Jr.

p. 143 The two men talked ... and details of EchoStar/Advanced negotiations and subsequent court case: compiled from interviews with Garner and Ergen; citations from the court file, including depositions, in *EchoStar Satellite Corp. vs. Advanced Communications Corp.*, civil case 94-N-2497, U.S. District Court, Colorado.

p. 143 An MCI executive later said ... : The source is William Welty, who owned a partial DBS slot and consulted for MCI. He sent a letter dated 10/23/95 to Garner identifying the MCI executive as Todd Olson. "As I told you before," Welty's letter ended, "I first wrote to MCI back when its founder Mr. McGowan was alive. I offered a strategic alliance with MCI at that time. My offer, like yours, was ignored."

p. 146 But Roberts did hook up ... : The primary source for the relationship between Roberts, Milken and Murdoch is an article by Greg Critser, *The Washington Post Magazine*, 6/16/96.

p. 147 "needed a nice, fat ..." and "Government is a reality ...": ibid.

p. 148 Reed Hundt's background, including "child of the sixties" and "I owe this job …": Ken Auletta, *The Highwaymen*, p. 194.

p. 148 "He was like Attila …": interview with James Quello.

p. 149 "I felt very strongly …": interview with Rachelle Chong.

p. 149 The agency began life … and the Hush-A-Phone example: Peter Huber, *Law and Disorder in Cyberspace: Abolish the FCC and Let Common Law Rule the Telecosm* (Oxford University Press, 1997), preface.

p. 149 "Reed was a passionate guy …": interview with Judy Harris.

p. 150 "Later last week …": transcript of Hundt's remarks on 2/28/94, as prepared for delivery, from the archives at www.fcc.gov.

p. 150 "Soviet-style regulation": Paul Farhi, *The Washington Post*, 5/1/94.

p. 150 It came in a July 1994 *Wired* magazine interview … and dialogue: David Kline, *Wired*, 7/94; used with permission.

p. 151 "I think he thought it was …": interview with Chong.

p. 151 Quello staffers set up a mock shrine … : interviews with Quello and Pete Belvin, a former Quello staffer.

p. 151 He prized a photograph … : The photo is featured inside the FCC's annual report for fiscal year 1995.

p. 151 "The FCC has become …": *The Denver Post*, 8/12/95.

p. 152 "one of the greatest scams …": Bryan Gruley, *The Wall Street Journal*, 3/17/97.

p. 152 Among Hundt's actions … and details of Scott Blake Harris's tenure at the FCC: public reports of the agency's activities and an interview with Harris.

p. 153 "I simply looked …": interview with Harris.

p. 153 "The failure to get the 110 slot …": interview with Robert Thomson.

p. 154 TCI as "the archenemy" … and "Primestar is not today …": Chris McConnell, *Broadcasting & Cable*, 6/12/95.

p. 154 "As Chief of Staff …": letter dated 8/15/95 from Betsey Wright to Reed Hundt.

p. 155 "The fact is, this senator …": *Congressional Record*, 9/29/95.

p. 155 Before the vote … : interview with Thomson.

p. 155 The two men discussed … : Hundt interview with Bob Scherman, *Satellite Business News*, 10/25/95.

p. 155 "Would this same …": Lloyd Covens, *DBS Digest*, 10/16/95.

p. 155 "MCI reaffirms its commitment …": letter dated 10/10/95 from Gerald H. Taylor to Hundt.

p. 155 "If poor Mr. Garner ..." and "A year from now ...": Hundt interview with Scherman, *Satellite Business News*, 10/25/95.

p. 156 "For once, the millionaires ...": Edmund L. Andrews, *The New York Times*, 10/14/95.

p. 156 "The majority gives ...": Quello's dissenting statement.

p. 156 "If you get in bed ...": interview with Garner.

p. 156 "If we had opted to bid ...": *The Denver Post*, 1/23/96.

p. 157 Three parties showed up ... and the auction scene: reconstructed from interviews with participants and news accounts, including *Bloomberg News* service updates on the auction rounds.

p. 157 "The American consumer ...": Liza McDonald, *Bloomberg News*, 1/24/96.

p. 157 "Are you sure you really ...": Greg Critser, *The Washington Post Magazine*, 6/16/96.

p. 158 "We weren't going to bid ..." and "There is no business model ...": *The Denver Post*, 1/25/96.

p. 158 "We felt it was worth ...": interview with Charlie Ergen.

p. 158 "We went into this auction ..." and "I think we're moving ..." and "If a license ...": *Bloomberg News*, 1/26/96.

p. 159 "All I can say is ...": Andrea Harter, *Arkansas Democrat-Gazette*, 1/27/96.

p. 159 "I always say ...": transcript of Hundt's remarks on 2/15/96, as prepared for delivery, from the archives at www.fcc.gov.

p. 160 "There is some concern ...": letter to Hundt dated 7/1/96.

p. 160 "It is something that happens ...": *Bloomberg News*, 7/3/96.

p. 160 "Nothing can stifle the growth ...": Harris's article at www.harriswiltshire.com.

p. 160 Murdoch and Roberts picked up shovels ... : Price Colman, *Broadcasting & Cable*, 10/28/96.

p. 160 "We went in prepared ...": interview with Ergen.

Chapter 9: Elvis Is in the Building

p. 161 It was all seven or ... : Melinda Gipson, *Satellite Business News*, 12/14/92.

p. 161 "In business school ...": David Hartshorn, *TVRO Dealer*, 4/94.

p. 162 "I thought, 'If these ... '" and subsequent quotes from Carl Vogel,

Charlie Ergen and other EchoStar officials: author interviews and reporting, except where indicated.

p. 163 "I just look at what's the cheapest …": Gipson, *Satellite Business News*, 12/14/92.

p. 164 Three of six Long March 2E … : EchoStar filings with the U.S. Securities and Exchange Commission, prior to the *EchoStar I* launch.

p. 165 On February 15, 1996 … and subsequent details of federal investigation: U.S. House of Representatives Select Committee on U.S. National Security and Military/Commercial Concerns with the People's Republic of China, known as the Cox Committee report, released in May of 1999. See specifically the section "U.S. Export Policy Toward the PRC." Also see Jeff Gerth and David E. Sanger, *The New York Times*, 5/17/98, and Eric Schmitt, *The New York Times*, 7/9/98.

p. 166 Three years later, under intense … : Jeff Gerth and David E. Sanger, *The New York Times*, 2/23/99.

p. 167 "Had the launch failed …": interview with Ergen.

p. 168 Four officials, including … : *The Denver Post*, 10/31/96.

p. 169 Ameristar, an independent … : author's notes, some of which appeared in *The Denver Post*, 6/17/96.

p. 170 "We look at our service …": *The Denver Post*, 6/12/96.

p. 170 "We said, 'We're Hughes … '": interview with Eddy Hartenstein.

p. 170 "I wouldn't say …": *The Denver Post*, 6/12/96.

p. 171 "If this were the Old West …": Jim Carrier, *The Denver Post*, 9/22/96.

p. 171 "We're making it cheaper …": *The Denver Post*, 9/11/96.

p. 172 "The first version was …": transcript of speech by Decker Anstrom to the Society of Cable Television Engineers in Nashville, Tennessee, 6/10/96.

Chapter 10: Angry Eyeballs

p. 173 "One of the issues …": transcript of John Malone's speech, 1/11/96.

p. 175 "The most powerful man …": *Vanity Fair*, 10/98.

p. 175 Gates reportedly told … : recounted by David Warsh, *The Boston Globe*, 2/19/95.

p. 176 "They've got to …": interview with Malone, some of which appeared in *Wired*, 6/98.

p. 176 Founder Steve Case had ... ; Case faxed ... ; and subsequent Gates quote and Case quote: Kara Swisher, *aol.com: How Steve Case Beat Bill Gates, Nailed the Netheads, and Made Millions in the War for the Web* (Times Books, 1998), prologue, pp. 104–105.

p. 176 "It was a mistake. ...": interview with Malone.

p. 177 Malone's attitude ... : interview with John Perry Barlow.

p. 177 Doerr's connection to Malone ... and "One-way cable TV ...": George Gilder, *Forbes ASAP*, 4/95.

p. 178 "It's great if you can ...": Malone remarks at Western Cable Show, 12/12/96.

p. 178 "10 opportunities a week ...": *The Denver Post*, 4/3/96.

p. 179 "You're living a Web lifestyle ...": *Nerds 2.01: A Brief History of the Internet*, PBS documentary, 11/98.

p. 179 Malone's doctor told him ... : Ken Auletta, *The Highwaymen*, p. 42.

p. 180 Theodore John Kaczynski ... and quotes from the Unabomber manifesto and case: from archives at www.courttv.com.

p. 181 "Virtually all my wealth ...": *The Denver Post*, 4/3/96.

p. 181 "Brendan is basically ...": Auletta, *The Highwaymen*, p. 37.

p. 182 "This is going to be ..." and subsequent dialogue: author's notes of the event.

p. 183 After he had been diagnosed ... and subsequent dialogue: TCI videotaped interview with Bob Magness.

p. 184 "We do not recall ...": *Multichannel News*, 8/19/96.

p. 184 "Now I won't ...": *The Denver Post*, 8/16/96.

p. 184 "the seven dwarfs" and "Rumors that I have ...": Price Colman, *Broadcasting & Cable*, 10/28/96.

p. 185 "The issue is simple ...": Gary Rivlin, www.Upside.com, 9/1/96.

p. 185 "They haven't met ...": Eben Shapiro and David Kirkpatrick, *The Wall Street Journal*, 10/25/96.

p. 185 He had been treated ... and "Girl, you'll figure ...": Henry Dubroff, *The Denver Business Journal*, 1/16/98.

p. 186 Before he died ... : *The Denver Post*, 6/26/97.

p. 186 After the funeral ... and "Nobody's watching ...": Dubroff, *The Denver Business Journal*, 1/16/98.

p. 186 He told Sparkman ... : interview with J. C. Sparkman.

p. 186 "The story gets a little old ..." and subsequent quotes: author's notes, some of which appeared in *The Denver Post*, 12/13/96.

p. 187 Six weeks later ... and subsequent details of programming conflicts: author's notes and research, some of which appeared in *The Denver Post*, 1/17/97, 1/21/97, 1/23/97 and 2/16/97.

p. 188 "We are not treated ...": Ed Quillen, *The Denver Post*, 2/18/97; used with permission.

p. 189 During a panel discussion ... : transcribed from C-SPAN, 2/97.

p. 190 "I'm a businessman.": Auletta, *The Highwaymen*, p. 115.

p. 190 "I had an absolute ...": Elizabeth Lesly, Gail DeGeorge and Ronald Grover, *Business Week*, 3/3/97.

p. 192 "We're a little bit like the French ...": transcribed from a videotape of the Western Cable Show, 12/11/96.

p. 192 "When the fox said ...": *The Denver Post*, 4/10/96.

p. 192 He worked in a building ... and subsequent reporting: author's notes of 2/11/97 interview with Peter Barton, some of which appeared in *The Denver Post*, 4/3/97.

p. 194 "This is obviously ..." and details of actions by Roy Bliss Jr.: interview with Bliss, some of which appeared in *The Denver Post*, 11/26/96.

p. 195 Across town ... : interview with Kenneth Gibson, some of which appeared in *The Denver Post*, 2/16/97.

p. 196 "The only way ...": Raymond Snoddy, *Financial Times* of London, 1/27/97.

p. 196 Around this time ... and "Rupert told us ...": interview with Barton.

Chapter 11: Calling Dr. Kevorkian

p. 197 Charlie Ergen's dream ... and subsequent reporting: author's notes, some of which appeared in *The Denver Post*, 3/30/97.

p. 199 "The Dish Network brand ...": Greg Tarr, www.e-town.com, 4/7/97.

p. 200 "There's nothing the phone ...": David Kline, *Wired*, 7/94.

p. 201 "There's no one who can say ...": *The Denver Post*, 3/1/97.

p. 201 "Simply stated ..." and "competing economic interests ...": Anne Marie Squeo and Bob Drummond, *Bloomberg News*, 3/31/97.

p. 202 "cynical attempt ...": *Satellite Business News*, 4/11/97.

p. 202 Leo J. Hindery Jr. arrived ... and subsequent material, except where indicated: author's reporting, interviews with Hindery and biographical material supplied by Hindery, some of which appeared in *The Denver Post*, 2/13/97.

p. 204 "I started dialing ...": Jennifer Reese, *Stanford Business*, 3/98.

p. 204 "I know people. ...": Geraldine Fabrikant, *The New York Times*, 6/21/98.

p. 204 Malone "needed me ..." and "TCI was a gas station company ...": Reese, *Stanford Business*, 3/98.

p. 205 "The company needs ...": interview with Gordon Crawford.

p. 206 "It's a damn dangerous game.": *The Denver Post*, 3/4/97.

p. 206 A cartoon ... : Leo Michael, *Multichannel News*, 4/28/97.

p. 206 "I have no problem ...": *Satellite Business News*, 3/19/97.

p. 206 "Let's not whale ...": Kim Mitchell, *Cable World*, 3/24/97.

p. 206 While chasing Sky ... : Sallie Hofmeister, *Los Angeles Times*, 3/28/97.

p. 207 Malone's Liberty Media considered ... and "There's a difference. ...": interviews with Peter Barton.

p. 207 "I certainly did ...": interview with John Malone, part of which appeared in *The Denver Post*, 5/17/98.

p. 208 "All roads lead ...": interview with Tim Robertson.

p. 208 "The up-front costs ..." and subsequent Rupert Murdoch remarks: transcript of his statement on 4/10/97.

p. 209 industry officials met ... : Steve Hamm, *Business Week*, 2/2/98.

p. 209 "We were meeting ...": transcript of Gates's interview on the *Charlie Rose Show*, 3/4/98.

p. 210 "For four hours I sat there ...": Ken Auletta, *The Highwaymen*, p. 289.

p. 210 "We think Asia ..." and dialogue: ibid., p. 265; used with permission.

p. 211 "Instead of teaming ...": David Lieberman, *USA Today*, 4/11/97.

p. 211 "It is unclear ...": Bob Scherman, *Satellite Business News*, 4/11/97.

p. 211 Remarkably, Murdoch left the message ... and Ergen got a call from Ted Turner: interviews with Ergen.

p. 211 "Are you suggesting ...": *The Denver Post*, 4/12/97.

p. 212 They debated the cost ... : John Lippman, Mark Robichaux and Bryan Gruley, *The Wall Street Journal*, 5/15/97.

p. 212 Padden walked out ... : interviews with David Moskowitz and Steve Schaver, EchoStar executives.

p. 212 EchoStar and News Corp. had planned … : filings in *EchoStar Communications Corp. vs. News Corp.*, civil case 97-960, U.S. District Court, Colorado.

p. 212 "The EchoStar deal …": Richard Zoglin, *Time*, 5/12/97.

p. 213 The Cato gala … : Roxanne Roberts, *The Washington Post*, 5/2/97.

p. 213 At the Oxford, Malone … : an event attended by the author, "the reporter," who questioned Malone.

p. 214 He reportedly told Murdoch early on … : Lippman, Robichaux and Gruley, *The Wall Street Journal*, 5/15/97.

p. 214 Meeting with Ergen and Schaver … : *EchoStar Communications Corp. vs. News Corp.*

p. 214 "The court clerk …": *The Denver Post*, 5/13/97.

Chapter 12: Summer of Love

p. 215 Ergen received condolence … and subsequent quotes: interview with Charlie Ergen and author's notes of press call.

p. 217 "Time Warner is not …": Kent Gibbons, *Multichannel News*, 5/19/97.

p. 217 In New Hampshire … : Kent Gibbons, Through the Wire column, *Multichannel News*, 5/19/97.

p. 217 Meeting on May 19 … : *The Denver Post*, 5/22/97.

p. 218 "Some days, you wake up …": Kent Gibbons, *Multichannel News*, 6/16/97.

p. 219 In June of 1997 … : Raymond Snoddy, *Financial Times* of London, 6/23/97.

p. 219 "The computer industry is converging …": George Gilder, *Forbes ASAP*, 2/23/94.

p. 219 "I am very worried …" and "We don't want Gates …": Snoddy, *Financial Times* of London, 6/23/97.

p. 220 Ergen sat in a high wingback chair … and subsequent reporting from the DBS event: author's notes, some of which appeared in *The Denver Post*, 6/12/97.

p. 221 "It's a win …": interview with John Malone.

p. 222 "EchoStar was a black hole …": Seth Schiesel, *The New York Times*, 6/12/97.

p. 223 "What was TCI's role …" and Malone's response: Chuck Ross, *Advertising Age*, 9/29/97.

p. 223 Tim Robertson, The Family Channel … : interview with Robertson.

p. 223 "We were going to be in opposition …": Snoddy, *Financial Times* of London, 6/23/97.

p. 224 One consequence of the deal … : Susan Whitney, *Deseret News*, 8/31/97.

p. 224 The core of the deal … : *The Denver Post*, 6/18/97.

p. 224 The deal was all set … and subsequent quotes from the participants, except where indicated: filings in the Magness estate civil case, 96PR944, district court, Arapahoe County, state of Colorado.

p. 225 Ritchie said he and Fisher … : interview with Dan Ritchie.

p. 225 But at 6:40 that evening … : The time-stamped fax is part of the court file.

p. 226 "It was a very strange time …": Kathryn Harris, *Los Angeles Times*, 4/17/94.

p. 226 "I was too weak. …": ibid.

p. 226 It was there he learned … : David Lieberman, *USA Today*, 5/8/95.

p. 227 During a flight to Japan … : Amy Barrett, Steve Hamm and Ronald Grover, *Business Week*, 9/21/98.

p. 227 QVC represented dollar democracy … and the Von Furstenberg episode … ; and the description of Diller's arrival at QVC: Ken Auletta, *The Highwaymen*, pp. 20–21.

p. 228 "They won. We lost. …": recounted by John Lippman, *Los Angeles Times*, 7/14/94.

p. 228 Comcast feared a loss … : Kathryn Harris and John Lippman, *Los Angeles Times*, 7/13/94.

p. 228 "captain of the cable team": Auletta, *The Highwaymen*, p. 26.

p. 228 In the first half of 1997, Ralph Roberts … : Joe Estrella, *Multichannel News*, 5/19/97.

p. 229 "What angered me …": Chuck Ross, *Advertising Age*, 12/22/97.

p. 229 "It's like trying …": interview with Peter Barton.

p. 230 "Hindery and TCI's New Spirit …": *The Denver Post*, 7/21/97; used with permission.

p. 232 "I don't know why …": interview with Malone.

p. 232 "We are happy …": Michael Blood, Associated Press, 7/24/97.

p. 232 "What Martin Luther King ...": Marianne Paskowski and R. Thomas Umstead, *Multichannel News*, 9/8/97.

p. 232 *Forbes* magazine published ... and listings: 10/13/97 issue; copyright 1997, *Forbes* magazine. Reprinted with permission.

p. 234 "As this committee deliberates ...": Laura Maggi, States News Service, reported in *Multichannel News*, 10/27/99.

p. 234 "I've been a very successful ..." and Quello reaction: interview with Quello.

p. 234 A book by Peter Huber ... : *Law and Disorder in Cyberspace: Abolish the FCC and Let Common Law Rule the Telecosm* (Oxford University Press, 1997); the quote is from the book's dust jacket.

p. 235 Charlie Ergen walked ... and subsequent reporting: author's notes, some of which appeared in *The Denver Post*, 10/5/97, 10/6/97 and 10/7/97.

Chapter 13: Endgame

p. 239 "After Rupert Murdoch ...": Consumers Union press release, 9/23/97.

p. 240 The force behind ... and Gene Kimmelman's background: interview with Kimmelman.

p. 241 "Certain animals compete ...": László Mérő, *Moral Calculations*, pp. 10–11; used with permission.

p. 242 "None of us had the sense ...": author's notes of the groundbreaking.

p. 243 "Intel can now do MPEG-2 ...": notes by Leslie Ellis, *Multichannel News;* used with permission.

p. 243 "The security issue ...": *The Denver Post*, 12/19/97. Robert J. Cleary, a first assistant U.S. Attorney in New Jersey and lead prosecutor in the Unabomber case, said in a 1999 interview that he could not discuss the potential targets of the Unabomber, but said that there was "absolutely no evidence of him sending an e-mail to anyone."

p. 244 "We are greatly concerned ...": joint press release by TCI and Comcast, 12/23/97.

p. 244 "Let's talk about Oprah ...": *The Denver Post*, 9/24/97.

p. 245 Hindery was so apoplectic ... : *The Denver Post*, 1/18/98.

p. 245 "I didn't give Mr. Garnett ...": Ted Hearn, *Multichannel News*, 10/13/97.

p. 245 Growing up in Detroit … : C. Michael Armstrong's personal and business background from AT&T, *Who's Who* listing and ABCNews.com biography.

p. 247 "AT&T was in here …": notes of 1998 interview with Malone by John Higgins and Price Colman, *Broadcasting & Cable;* used with permission.

p. 248 He filed against MCI … and subsequent details of lawsuit: *Advanced Communications Corp. vs. MCI Communications Corp.,* civil case LR-C-98-218, U.S. District Court, Eastern District of Arkansas, Western Division.

p. 248 "The Advanced case …": Stephen Keating, *Cablevision,* 2/22/99.

p. 249 Michael Milken settled … : Peter Truell, *The New York Times,* 2/27/98.

p. 249 Milken's most enduring legacy … : Floyd Norris, *The New York Times,* 3/1/98.

p. 250 A holiday television ad … : *The Denver Post,* 12/12/97.

p. 250 "I told him …" and subsequent conversation with Carl Vogel: interview with John Reardon.

p. 251 "There's no need …" and "A good friend of Bill's …": Associated Press, 5/4/98.

p. 251 the cable gang pulled out $479 million … : U.S. Securities and Exchange Commission filings by Primestar.

p. 252 He grew up … : Jeffrey H. Birnbaum, *Fortune,* 6/8/98.

p. 252 Klein was criticized … : Catherine Yang, *Business Week,* 5/26/97.

p. 252 "surgical intervention": Michael Hirsh, *Newsweek,* 3/9/98.

p. 252 One measure of market … : Federal Trade Commission merger review guidelines.

p. 252 The deal "had a chilling …" and subsequent quotes: *United States of America vs. Primestar Inc., et al.,* civil case 1:98CV01193, U.S. District Court, District of Columbia.

p. 253 "It was a little bit …": Keating, *Cablevision,* 2/22/99.

p. 254 Glenn Jones enjoyed … : interview with Jones, and author's attendance at show.

p. 254 "Aim at that satellite …": lyrics by Erica Stull; used with permission.

p. 255 A poem titled … : Glenn Jones, *Briefcase Poetry of Yankee Jones,* vol. 3 (Glenn R. Jones, 1985), p. 44; used with permission.

p. 255 Over the Memorial Day … : Tom Kerver, *Cablevision,* 6/8/98.

p. 256 Allen's Vulcan Ventures ... : Elizabeth Lesly and Kathy Rebello, *Business Week,* 11/18/96.

p. 256 Allen was techie ... : ibid.

p. 257 In early June, TCI ... : Kent Gibbons, *Multichannel News,* 6/15/98.

p. 257 "Manchurian candidate" ... : transcript of Subcommittee on Communications, U.S. Senate Committee on Commerce, Science and Transportation, 11/16/89, p. 176.

p. 258 "Two or three companies ...": Doug Halonen, *Electronic Media,* 1/31/94.

p. 258 In Milford, Connecticut ... : *The Denver Post,* 6/28/98.

p. 258 "Malone piles on debt. ...": James J. Cramer, *TheStreet.com,* 6/26/98; used with permission.

p. 259 "TCI has made most ...": John Higgins/Price Colman interview notes with Malone; used with permission.

p. 260 "You get more bang ...": interview with Malone.

p. 261 When the J. D. Power ... : The survey was of 11,000 subscribers; Primestar, EchoStar and DirecTV led the ratings.

p. 262 In June of 1998 ... and details of EchoStar/News Corp. settlement: author's reporting and interviews with Ergen, some of which appeared in *The Denver Post,* 12/1/98.

p. 263 Evident at the trial ... : author's reporting at the trial, some of which appeared in *The Denver Post,* 3/14/96, 3/16/96 and 3/23/96.

p. 264 Carl Vogel sat ... and subsequent details: author's reporting and interview with Vogel, some of which appeared in *Cablevision,* 2/22/99.

Epilogue: Survival of the Leveraged

p. 267 "It's the Wall Street way. ...": Dunstan Prial, *Dow Jones Newswires,* 3/29/99.

p. 267 Rupert Murdoch dodging ... : *The Economist,* 3/20/99. See also Paul Farhi, *The Washington Post,* 12/7/97.

p. 267 Ever vigilant, Ted Turner ... and quotes: Kent Gibbons, *Multichannel News,* 5/11/98.

p. 268 "Rupert Murdoch is feeling ...": Henry Porter, *The Guardian,* 9/10/98.

p. 268 Murdoch returned to Adelaide ... : Reuters News Service, 11/16/98.

p. 268 Yet, Murdoch nixed plans ... : *The New York Times*, 9/14/98.

p. 268 In 1993, he told News Corp. advertisers ... : Ken Auletta, *The Highwaymen*, p. 267.

p. 269 "We're not proud ...": ibid., p. 268.

p. 269 Murdoch reportedly spiked ... : *Daily Telegraph* of London, 2/27/98.

p. 269 Jiang praised Murdoch ... : *People's Daily* of China, 12/11/98.

p. 269 The much larger Herald Group ... : Thomas Kiernan, *Citizen Murdoch*, pp. 46–47.

p. 269 "I then did something ...": Marc Gunther, *Fortune*, 10/26/98.

p. 270 His opponent did indeed ... : Kiernan, *Citizen Murdoch*, pp. 46–47.

p. 270 "I was very depressed ...": Raymond Snoddy, *Financial Times* of London, 6/23/97.

p. 271 In March of 1999, Comcast ...: details from Price Colman, *Broadcasting & Cable*, 5/10/99.

p. 272 studied thermodynamics ... : transcript of Subcommittee on Communications, U.S. Senate Committee on Commerce, Science and Transportation, 11/16/89, p. 108.

p. 273 "You've got to be hedged ...": notes of press conference.

p. 274 "Is this a deposition? ..." and remaining quotes: interview with Ergen.

ACKNOWLEDGMENTS

T HIS BOOK IS MY interpretation of individuals, events, strategies and deceptions, based on interviews, public records, news accounts and other research.

It would not have been possible without the information and insights provided by a host of people, including all those mentioned in the text, those cited in the source notes and several people who asked to remain anonymous. A primary source for this book was my reporting on the cable and satellite TV industries for *The Denver Post* between mid-1995 and mid-1998, before taking a one-year leave of absence. I am thankful to the newspaper for those opportunities.

I am much obliged to Tom Kerver of *Cablevision* magazine and Bob Scherman of *Satellite Business News*, who were generous with their time and recollections. Special *menschen* to Maureen Harrington and Steven Wilmsen, both of whom supplied bright ideas, sage advice and encouragement. Her wisdom: "Maxwell Perkins is dead." His: "Come to the table when dinner is served."

Thanks also to my dear friends: Anya Breitenbach, Patricia Callahan, Matthew Carden, Scott Gerlock and Martha Rasmussen.

My appreciation to the folks at Intellinet, who provided baseline technical assistance; to the Wesleyan Writers Conference for its support; to Bill Long, a newspaper tiger; and to whomever invented the Internet.

Cheers to Laurie Harper at the Sebastian Agency in San Francisco, who believed in *Cutthroat* before it existed, and to the crew at Johnson Books in Boulder, who published it once it did.

And dearly, to Jack, Meems, Boo, Christopher, Tom, the Arch and Lynn.

Stephen Keating
Denver, Colorado
August, 1999

305

INDEX

ABC, 2, 15, 36, 97, 142, 198, 207, 212;
  cable companies and, 201
A. B. Hirschfeld Press: Antares and, 88
Action for Children's Television Lecture on Media and Children, 150
Adelaide *News*, 15, 269
Adelaide *Sunday Mail*, 269
Advanced Communications Corp.,
  103, 141, 155, 248–49; application
  by, 153–54; petition by, 116;
  TCI/DBS and, 145
Advertiser: Murdoch and, 270
Advertising Age: Malone in, 223
Ailes, Roger, 18, 20
Albert, Carl, 47
Alcoa, 110, 113
Alden, Paul, 73, 74
Ali, Muhammad, 24, 47
Allen, Paul, 24, 122; cable television
  and, 271; Gates and, 256; investments of, 256; wealth of, 233
Allen, Robert, 245
Allen & Company, 70, 192
Alpert, Mickey: on Murdoch, 221
AlphaStar, 217
Amazon.com, 259
American Express, 53
American Movie Classics, 99
American Sky Broadcasting (ASkyB),
  13, 21, 26, 28, 158, 212; groundbreaking for, 160; Murdoch and,
  160; Primestar and, 211
American Telephone and Telegraph
  (AT&T), 37, 110, 200; Armstrong

and, 246–48; Baby Bells and, 247,
  258, 274; breakup of, 56; cable television and, 271; Ergen and, 274;
  Liberty Media and, 272, 273; Malone and, 61, 247, 248, 257; MCI
  and, 147; MediaOne and, 272; satellites and, 47, 109; TCI and, 63,
  257–58, 259, 261; Teleport and, 248
American Television & Communications Corp. (ATC), 53, 54
America Online (AOL), 30, 176, 272;
  DirecTV and, 265
Ameristar, 169
Ameritech, 135, 200, 247; cable rates
  and, 240
Andreae, Fred: on Malone, 60
Andreessen, Marc, 179
Animal Planet, 189, 190, 191, 196
Annunziata, Robert, 247
Anschutz, Philip, 274
Ansett, loan for, 14
Anstrom, Decker, 172
Antares Satellite Corp., 88, 89
Anthony, Barbara Cox: wealth of, 233
Antitrust actions/restrictions, 20, 42,
  56–57, 79, 104, 129–30, 177, 179,
  200, 251–52
AOL. *See* America Online
Apple, Gordon: DBS and, 140–41
Arianespace, 166, 168, 171
Arkansas Democrat-Gazette: Garner in,
  159
Arms Export Control Act: violation of,
  97